Me, Myself and Infrastructure
Private Lives and Public Works in America

GREGORY K. DREICER

with Tiffany Soule Anderson, Lauren DeMille, Emily Duwel,
Mary Mahon, Alex Marshall, Leili Towfigh

ASCE
AMERICAN SOCIETY OF CIVIL ENGINEERS

Chicken&Egg Public Projects

Library of Congress Cataloging-in-Publication Data

Me, myself and infrastructure: private lives and public works in America/Gregory K. Dreicer/ Chicken&Egg Public Projects
   p. cm.
   Includes bibliographical references and index.
   ISBN 0-7844-0611-1
      1. Public works—United States—Exhibitions.
      2. Civil engineering—United States—Exhibitions.
      I. Chicken&Egg Public Projects (Firm)

   TA23 .M42 2002   2002 018420
   624'.0973—dc21

The contents of this publication should not be construed to be a standard or endorsement of the American Society of Civil Engineers. ASCE makes no representation or warranty of any kind, whether express or implied, concerning the accuracy or completeness of any information.

ASCE and American Society of Civil Engineers—Registered in U.S. Patent and Trademark Office.

Photocopies: Authorization to photocopy material for internal or personal use under circumstances not falling within the fair use provisions of the Copyright Act is granted by ASCE to libraries and other users registered with the Copyright Clearance Center (CCC) Transactional Reporting Service, provided that the base fee of $8.00 per chapter plus $.50 per page is paid directly to CCC, 222 Rosewood Drive, Danvers, MA 01923. The identification for ASCE Books is 0-7844-0611-1/02/$8.00 + $.50 per page. Requests for special permission or bulk copying should be addressed to Permissions & Copyright Department, ASCE.

Copyright © 2002 by the American Society of Civil Engineers. All Rights Reserved.

ISBN 0-7844-0611-1

Manufactured in the United States of America.

Catalogue published in conjunction with the exhibition.

Me, Myself and Infrastructure:
Private Lives and Public Works in America

THE NEW-YORK HISTORICAL SOCIETY
(May 21, 2002 – September 15, 2002)

THE NEW YORK PUBLIC LIBRARY
SCIENCE, INDUSTRY AND BUSINESS LIBRARY
(May 21, 2002 – December 14, 2002)

NATIONAL BUILDING MUSEUM
(October 1, 2002 – February 2, 2003)

The exhibition will travel to an additional venue in 2003.

The American Society of Civil Engineers (ASCE), underwriter of its 150th anniversary programs, represents more than 125,000 members of the civil engineering profession worldwide. Founded in 1852, it is America's oldest national engineering society. ASCE's vision is to position engineers as global leaders building a better quality of life. Its mission is to develop leadership, advance research, and advocate lifelong learning. For additional information, please visit www.asce.org.

Me, Myself and Infrastructure is part of the ASCE 150th anniversary celebration and is made possible by the support of the American Society of Civil Engineers Foundation, The Elizabeth & Stephen Bechtel, Jr. Foundation, and Charles Pankow Builders, Ltd. The ASCE Foundation is proud to sponsor products and programs that make a major contribution to technology education in the United States.

Chicken&Egg Public Projects™ creates interpretive environments and interactive strategies that advance public understanding of cultural and social issues. Our exhibition, print, and multimedia projects elicit powerful responses and build institutional identity. Cross-fertilization is fundamental to our creative approach. Chicken&Egg develops projects that integrate conception, development, fabrication, and dissemination.

www.chickenandegg.org

Exhibition design: Chicken&Egg Public Projects, Inc. and Boym Partners, Inc.

Catalogue design: Hall Smyth, Chicken&Egg Public Projects

6	PREFACE
7	FOREWORD
8	INTRODUCTION
9	CHAPTER GUIDE

10 — Who's Responsible?

34 — Is It Safe?

64 — Why So Big?

82 — Is It Available?

114 — How Much Does It Cost?

138 — How Long Will It Last?

162	AFTERWORD
164	ACKNOWLEGEMENTS
165	BIBLIOGRAPHY
168	CREDITS

Preface

CIVIL ENGINEERING touches practically every aspect of life. Highways for commuting, clean water for drinking, and buildings for working, living, and playing are just a few examples of civil engineering's vital role in our society's past, present, and future. Our nation's economy—in fact, the world's economy—is based on an infrastructure that undergirds our communities. Many conveniences would be impossible without the technology and infrastructure civil engineers provide.

Civil engineers offer the world solutions for today's problems and look ahead to tackle tomorrow's challenges. Earthquake codes were developed largely by civil engineers striving to improve public safety. Wind hazard codes are currently being researched and strengthened to ensure a safer quality of life for inhabitants of wind risk areas. Civil engineers examined the blast damage of the Murrah Federal Building in Oklahoma City to assure that new federal construction reflects the lessons learned from this tragedy. Currently, civil engineers are looking at the multitude of factors involved in the September 11th tragedies at the World Trade Center and the Pentagon, searching to make our world more secure in the face of new hazards.

As the world edges into a new century, and the American Society of Civil Engineers embarks on another 150 years, possibilities abound for the profession to play an integral role in solving the world's most pressing needs. From the crying need for modern transportation, power, water and wastewater systems in the developing world to the explorations and adventures in space, the challenges facing humanity are great. These challenges demand innovative civil engineering solutions. It is essential that civil engineers and the public develop an effective partnership to forge new, creative paths for the near and distant future.

Me, Myself and Infrastructure is an exciting engagement of civil engineering and the public. On behalf of the American Society of Civil Engineers, and as a civil engineer, I am excited to be a part of this innovative encounter between the civil engineering community and the public we serve. This can only be another step toward our ultimate goal of *Building a Better World*.

H. GERARD SCHWARTZ, JR., PH.D., P.E.
President, American Society of Civil Engineers

Me, Myself and Infrastructure
Private Lives and
Public Works in America

Property of Robert Collins
purchased in Chicago, Ill. AT THE
AMERICAN INST. of Architects STORE
2007
_____

PLEASE RETURN SOME DAY OR
PASS ON TO SOMEONE WHO WOULD
APPRCIATE.
       Thanks   RAC

Foreword

FEW AMERICANS realize that almost every aspect of their existence depends upon an infrastructure that makes the conveniences of modern life available on a daily basis. Our lights, bathtubs, telephones, refrigerators, stoves, computers, and television sets, to mention only a few obvious examples, would go dark and dry if our power and water systems went down. Indeed, so dependent have we become on modern infrastructure that even when we go camping we are rarely disengaged from the complex engineering marvels that define our civilization.

What holds true for Americans in general is doubly important for the citizens of New York City, the most densely settled and complex human environment in North America. Nowhere else on earth do so many hundreds of thousands of people both live and work in tall buildings. Nowhere else do so many electrical, sewer, telephone, and transportation systems intersect in such restricted spaces. Nowhere else is there gridlock even on the sidewalks.

But the importance of Gotham goes beyond its astonishing complexity; the city encompasses the history of civil engineering. In 1852, the first meeting of what would become the American Society of Civil Engineers took place in the New York City office of the Croton Aqueduct, a model of urban water supply. Manhattan gave birth to America's first elevators, its first elevated railroads, its first telephones, and its first skyscrapers. It was here that the first public power station began the electrical illumination of America. In New York, engineers developed infrastructure and put it to the ultimate test.

This marvelous exhibition and book remind us of how much we depend upon the dedication and expertise of the civil engineers who make possible the lifestyles that we cherish.

KENNETH T. JACKSON
Director, The New-York Historical Society

From facing page left to below right:

Engineers at the Croton Aqueduct, N.Y.
Title page of 1942 book on civil engineering as a career
Computerization, Department of Traffic, N.Y., 1964
Construction truck, Army engineers, Fort Knox, Kentucky, 1942
Engineer and "largest map in the world," Wellesley, Mass., 1930
Sears, Roebuck, and Co. catalogue, 1940
Sewage purification plant, Columbus, Ohio
Stuck in the mud, Virginia, 1909
Construction of Memorial Bridge, Washington, D.C.
Boy pumping water, Brooklyn, N.Y., 1898

Introduction

Me, Myself and Infrastructure: Private Lives and Public Works in America tells the story of a thirsty, car-crazy nation. This project explores the relationship of the public to the civil engineering networks that define modern life.

Your relationship to civil engineering networks, also known as *infrastructure* or *public works,* is private. Uncovering this relationship reveals your beliefs about how you want to live and what you think of the people in your neighborhood. It exposes the intensity of your concern for the environment. It demonstrates what you know about the history of everything around you.

Civil engineers are the technologists behind an infrastructure that includes buildings and bridges; water supply and treatment systems; energy, transportation, communication, and aerospace networks; life safety measures; and land use planning. Civil engineers are the innovative problem solvers behind research, design, management, and the formulation of public policy.

The location of your home, the visibility of a crosswalk, the taste of water—your values and behavior determine these. At the same time, your home, your streets, the coffee in your cup—these shape you.

One-third of the world's population lacks access to safe drinking water. One-half lack the sanitation systems that keep things flush in the United States. You'd think that these facts would lead Americans to worship their civil engineers. Wrong. Why? Civil engineers do their job too well. They make it seem too easy. They make it seem natural.

Me, Myself and Infrastructure examines the indivisible links between our private lives and public works.

Dedication of Red Jacket Bridge, Minnesota, 1911

Chapter Guide

Each chapter of *Me, Myself and Infrastructure* explores the fundamental questions behind infrastructure and your role in its construction. The answers define your life.

Who's responsible?

introduces the people behind suburban life in Washington, D.C.

Parents
President
Cyclists
Automotive Entrepreneur
Paving Innovator
Housing Policymaker
Highway Administrator
Civil Engineering Leaders
Real Estate Developer
Fast Food Founder
Chain Store Mogul
County Official
U.S. Senator
State Governor
Transportation Engineer
Web Entrepreneur
New Homeowners

Is it Safe?

crosses streets in Atlanta, New York, and Portland, Oregon, in search of the origins of security.

Safety Today
Construction of Safety
Future Safety

Why so big?

examines the link between home and infrastructure.

Community
Structure
Infrastructure
Backyard

Is it available?

analyzes the essential ingredient of every coffee shop in New York.

Convenience
Dependability
Quality
Safety
Sustainability
Cost
Security
Community
Disposal and Treatment

How much does it cost?

explores the underpinnings of the information age in a Silicon Valley office building.

Place
Manufacture
Use
Disposal

How long will it last?

reflects on the durability and disposability of materials and structures.

Permanent
Impermanent
Gone
Forever

Photographs

New residential subdivision, Fairfax County, Virginia

Who's Responsible?

Parents	14
President	15
Cyclists	17
Automotive Entrepreneur	18
Paving Innovator	19
Housing Policymaker	20
Highway Administrator	22
Civil Engineers	23
Real Estate Developer	24
Fast Food Founder	25
Chain Store Mogul	26
County Official	27
U.S. Senator	28
State Governor	28
Transportation Engineer	29
Web Entrepreneur	30
New Homeowners	32

An affordable house, yard with fresh air, nice neighbors, good schools nearby, easy transport…

Americans of all backgrounds have been told, and many believe, that this is their dream. As a result, more than half of the population lives a suburban life.

During the last 50 years, federal, state, and local governments funded the construction of thousands of miles of roads. This enabled the building of the residential subdivision, the shopping center, and the office park. Together, these made a new world: the modern suburb. The creation of its infrastructure—roads, water, sewerage, gas, power, telephone, and Internet—is *the* monumental technological and cultural achievement of 20th-century America. Infrastructure frames every snapshot; it holds the memory and future of all Americans.

Across the U.S., suburbs continue to expand more rapidly than the cities they surround and upon which they still depend. After New York, Los Angeles, and Chicago, the Washington, D.C. metropolitan area is the most populous in the United States: 7.6 million people, quite a few of whom get stuck in traffic on the way to work. Who's responsible?

Mom and Dad PARENTS

Who else? The engineers of your life. The builders of your first networks—the ones that shaped you. Mom and Dad were around, or not around, and they supported you, or didn't. They're easy to blame—but are they responsible for the way you turned out?

You can't choose your parents any more than you can choose the infrastructure into which you are born. Make your parents change? Difficult. Alter the infrastructure? It's possible.

Both your parents and civil engineers are responsible for your survival—parents during your early years, civil engineers every day of your life. You hold both accountable for conditions they may not have created or could not foresee. The better they are, the easier they are to take for granted. Perhaps this is why parents and civil engineers are misunderstood, resented, and revered.

George Washington PRESIDENT

The District of Columbia is where it is because of George Washington. His home, Mount Vernon, was in Fairfax County, Virginia.

The debate over the location of a capital district, "not exceeding, ten miles square," according to the Constitution, dominated political debate for two years before the decision was made in 1791. The U.S. seat of government was placed within two slave-owning Southern states in exchange for bringing to the entire nation the commercial capitalism that characterized the North.

Although Georgetown, in D.C., and Alexandria, Virginia were then near the geographical center of the U.S., this was not the primary reason that the capital arose in land formerly owned by Virginia and Maryland. George Washington, who selected the site, with the help of Virginians Thomas Jefferson and James Madison, was deeply involved in his plantation. He believed that nearby Alexandria could compete with Philadelphia and Baltimore.

From 1791 until his death in 1799, George Washington's main focus was the establishment of the capital. Engineer and architect Pierre Charles L'Enfant, who produced the plans for the District of Columbia, told Washington that the plan had to allow for expansion. The federal government, however, never used the Virginia land across the Potomac and it was returned to the state in 1846.

Mount Vernon, Fairfax County

Today only 11% of the population in the Washington, D.C. metropolitan area live within the District of Columbia. Most live in the suburbs, an expanding region of homes and businesses that surround the capital city. Its unifying feature is the 64-mile Beltway, I-95 and I-495, which encircles the capital.

Like most suburbs in the United States, the Washington suburbs are no longer subordinate. Although their growth was dependent on the District of Columbia, they have become more powerful, both economically and politically.

FAIRFAX COUNTY

Fairfax County encompasses 399 square miles of land across the Potomac from Washington, D.C. It includes a significant part of the Beltway and part of Dulles International Airport, which also lies in neighboring Loudoun County.

> "The paths of the pioneer have widened into broad highways. The forest clearing has expanded into affluent commonwealths. Let us see to it that the ideals of the pioneer in his log cabin shall enlarge into the spiritual life of a democracy where civic power shall dominate and utilize individual achievement for the common good."
>
> —Historian FREDERICK JACKSON TURNER, *Contributions of the West to American Democracy*, 1903

Fairfax has gone from tobacco farming region worked by slaves to postwar bedroom community to self-described "thriving urban county" with the headquarters of six Fortune 500 companies. Between 1970 and 2000, its population more than doubled. It is now about 965,000. In 2000, single families occupied 51% of its 358,960 housing units. The median family income is $95,000.

The Fairfax County Economic Development Authority has offices in Germany, Japan, and London. It claims that "Fairfax County is the Home of the Internet with more than 50 percent of Internet traffic worldwide passing through northern Virginia every day." Hundreds of computer-related companies are located in Fairfax and Loudoun counties.

The northern Virginia suburbs have become the Silicon Valley of the East. The presence of Dulles Airport, the Defense Department's involvement in the development of the Internet, and a large number of engineers once employed by the defense industry contributed to the growth of the area. After the government deregulated AT&T competitors such as MCI realized that they needed to be near decision-makers in Washington.

Above: Sallie Mae headquarters
Bottom: Nextel Communications headquarters

The Wheelmen   CYCLISTS

The bicycle, not the car, was behind the initial impetus for the paving of roads. Bicycles, which gained tremendous popularity in the 1880s and 1890s, were much easier to use on paved or gravel surfaces. The League of American Wheelmen, a national cyclist group, organized a campaign for better roads, which led to the creation of The Office of Road Inquiry under the Agriculture Department. Eventually renamed the Bureau of Public Roads, it would grow into the dominant institution behind road building. In 1967, it was renamed the Federal Highway Administration of the U.S. Department of Transportation.

Wheelmen, c. 1890

Henry Ford AUTOMOTIVE ENTREPRENEUR

Henry Ford founded the Ford Motor Company in 1903. His corporation became rich by mass-producing automobiles and charging less for them than anyone else. No longer was the car a luxury for the well-to-do. The Model T, introduced in 1908, was a model of simplicity and dependability, and as practical on the farm as in the city. In the early twenties, Ford Motor Company sold over half the cars in the U.S.; it was producing 9,000 cars a day. By 1927, when Ford discontinued the Model T, 26 million automobiles were registered, one car for every five citizens.

At the 1939 World's Fair, Ford and its competitor General Motors built the two most expensive exhibition buildings. Both featured "roads of tomorrow," highways that would allow Americans to realize the full potential of their automobiles. Norman Bel Geddes, who designed the GM pavilion, wrote, "although our cars have been designed for efficiency and economy, the loss due to traffic congestion in New York City alone is a million dollars a day."

Left: Colonial Beach, Virginia, 1913 *Below:* Model T

Prevost Hubbard
PAVING INNOVATOR

Prevost Hubbard was a chemist who worked on the development of asphalt paving materials beginning in 1905. Within the Bureau of Public Roads, and later, the Asphalt Institute, he helped develop paving standards for highway construction. Hubbard was educated at the George Washington University in Washington, D.C., where in the 1870s Pennsylvania Avenue was one of the first roads in the U.S. to be paved with asphalt. During the 1920s, Hubbard led an effort that reduced the types of asphalt cement from 88 to nine. Later he helped create the first standard test for measuring paving strength.

Despite pressure from bicyclists, and even after large numbers of people began buying cars, widespread paving of roads did not immediately occur. In 1922, 80% of the roads in the U.S. were dirt and gravel. Civil engineers, scientists, manufacturers, and builders worked on perfecting reliable, cost-effective, durable paving methods. Early paved roads were often water-based macadam—better than dirt and gravel, but nothing like today's roads.

"People from the centers of population are moving out into the adjacent territory buying lots and building homes on many of the unimproved roads. Especially is this noticeable in Fairfax County. After these homes have been built a demand is made for a road that will carry traffic through the year; for many of the owners work at distant points and motor daily to their work. This has created a situation that is requiring large expenditures to meet only a part of these demands. In fact, it was only possible to take care of very few of these roads during the past year, and the prospects for the coming year are not bright."

—*Report of The State Highway Commission to The Governor of Virginia*, 1935

Virginia roads, before and after   1907   1920   1926

Uncle Sam HOUSING POLICYMAKER

"No agency of the United States government has had a more pervasive and powerful impact on the American people over the past half-century than the Federal Housing Administration," wrote historian Kenneth T. Jackson. Established in 1934 to stimulate the construction industry and revitalized by the GI bill ten years later, the FHA insured banks against loss. This enabled them to provide loans to homebuyers who would have been considered too high a risk. For many citizens, it became cheaper to buy than to rent, and it was no longer necessary to save money to buy a house. America became suburban.

The policies of the Federal Housing Administration encouraged white Americans to exchange urban dwellings for suburban single-family homes, while ensuring that white and black Americans would not live together in new suburban neighborhoods. FHA-insured loans were available to residents of the new suburbs, but not to those left behind in the city.

Today, one of the largest black middle-class populations in the United States lives in and around Washington, D.C. While attitudes change, racial boundaries endure: 29% of the metropolitan area's African Americans live in D.C. and 42% live in neighboring Prince George's County, known as a "black suburb." Eleven percent live in Montgomery County; 7% live in Fairfax County; 4% in Prince William County; 2% in Arlington County; and 1% in Loudoun County.

In 1950, the population of Washington, D.C. was approaching one million. By 2000, it was 572,000. This astonishing drop is typical of many cities in the United States. The population of St. Louis, for example, went from almost 900,000 in 1950 to 350,000 today.

Was the decline of the city inevitable? Not at all. After World War II, federal, state, and local governments managed the highway boom. Soon, every city was ringed by a tax-payer funded network of roads that opened a new frontier to development. Urban governments neglected downtowns and pursued "urban renewal," demolishing buildings and streetscapes that today are recognized as the most appealing urban features. Urban infrastructure created for a dense population was neglected in favor of an infrastructure built to serve a population spread over millions of acres of land. The future built then is our present. It is only one of many worlds that might have been built.

This new form of civilization, the automobile-oriented suburb, brought blessings and challenges. The roads allow more Americans to live in single-family homes than in any nation on earth. The suburbs give families more privacy, space, and greenery. Suburbs are also sources of traffic congestion, pollution, and a car-centered life.

"Most postwar families were not free to choose among several residential alternatives. Because of public policies favoring the suburbs, only one possibility was economically feasible."

—Historian KENNETH T. JACKSON, *Crabgrass Frontier*, 1985

APPLY

CONSULT

PLAN

ENJOY

With a variety of marketing tools including pins, brochures, and technical bulletins the Federal Housing Administration promoted low cost loans.

Frank Turner HIGHWAY ADMINISTRATOR

Frank Turner, the civil engineer who served as head of the Bureau of Public Roads and its successor, the Federal Highway Administration, is considered the father of the U.S. Interstate System. An expert in road construction who received his civil engineering degree from Texas A&M, Turner worked in a variety of road construction, maintenance, and research positions before gaining national influence.

Turner's technical skills and political savvy enabled him to play a decisive role in the construction of America's highways. He assisted legislators who developed plans for the Interstate Highway System in 1956; he was then put in charge of implementing the plans. Turner believed that engineers, whose work appeared to be independent of politics, would be best suited to guide the construction of the system. In his view, highways were to function as transportation arteries, not as regulators of land use.

During the 1960s, increasing awareness of the impact of highway construction on displaced communities and natural landscapes led the public to delay and occasionally stop construction. This was a surprise to civil engineers who believed that they could somehow remove themselves from the politics of public works. Eventually, it became necessary to invite the participation of the communities whose lives would be affected and to pay closer attention to the environment.

The Federal Highway Act of 1921 established a framework that is still in place. The federal government works with state highway departments to build roads constructed under federal standards. This provides for federal leadership in the management and engineering of roadways.

"America lives on wheels, and we have to provide the highways to keep America living on wheels and keep the kind and form of life we want."

—Secretary of the Treasury GEORGE M. HUMPHREY, after the passing of the act that funded the Interstate Highway System, 1956

Below: Dulles Toll Road. *Right*: Frank Turner, President Eisenhower, and the Advisory Committee on the National Highway Program, 1955

Civil Engineers

PUBLIC LEADERS

Road building helps determine the structure of society. How many roads should be built, where should they be built, and who should they serve? Civil engineers have helped ask and answer these questions. They have debated whether they should be asking these questions, providing the answers—or both.

Civil engineers both take direction and provide it. They work in private industry, government, management, research, innovation, design, construction, and manufacture. Civil engineers are hardly of one voice. They are a diverse group and debate within the profession has always been strong on issues ranging from national policy to the design of a curve in the road. There are many answers to each question and each answer raises more questions.

The roads, along with the water, sewer, power, and communication systems, have enabled millions of people and hundreds of businesses to move to the Washington, D.C. suburbs. This infrastructure was put into place by a network of civil engineers, elected officials, real-estate developers, entrepreneurs, and individuals making choices for themselves and their families.

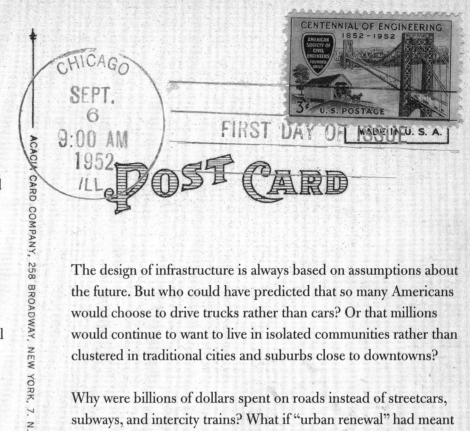

The design of infrastructure is always based on assumptions about the future. But who could have predicted that so many Americans would choose to drive trucks rather than cars? Or that millions would continue to want to live in isolated communities rather than clustered in traditional cities and suburbs close to downtowns?

Why were billions of dollars spent on roads instead of streetcars, subways, and intercity trains? What if "urban renewal" had meant the strengthening of urban infrastructure? The United States would be a different nation. The ingenuity and scale of the planning, design, and construction work managed by civil engineers is astonishing. The choices that drove civil engineers, however, were the choices of voters, taxpayers, and citizens.

John T. "Til" Hazel REAL ESTATE DEVELOPER

Lawyer and real estate developer John Hazel specializes in assembling large areas of land for residential and commercial use. Hazel's brother William is a building contractor specializing in clearing and leveling land, installing sewer pipes, and building roads.

In 1957, Harvard-trained Hazel began his first job: condemning rural land for a portion of an Interstate highway. This stretch of road, also part of the Beltway, ten miles outside of the District of Columbia, allowed the highway to run the length of the east coast. Tysons Corner, Virginia, was created by a triangle formed by the intersection of Routes 7, 123, and the Beltway.

Between 1950 and 1960, the population of Fairfax County tripled; it almost doubled in the next ten years. Hazel was the force behind the rezoning that allowed rural land to become the site of the shopping centers and suburban neighborhoods that make Fairfax one of the nation's wealthiest counties. Hazel won legal cases that stopped the mandating of affordable housing and the creation of a five-year plan of growth. Today, John Hazel attributes transportation difficulties to local communities that prevented the construction of additional highways.

In 2001, Gannett, the international news company, moved its headquarters and 1,700 employees from Rosslyn—located near a Metro station across the Potomac from Washington, D.C.—to new buildings in Tysons Corner.

"Put simply, Tysons Corner has all the same components of a major city, yet lacks any of the infrastructure needed to make it work," wrote J. Daniel Malouff in *BeyondDC.com*. This is one of the gentler critiques of Tysons, which contains millions of square feet of office space. Tysons was built for automobile traffic. Transforming it into a place where pedestrians can walk, or where a proposed mass transit line would function effectively, is an enormous challenge.

Tysons Corner, Fairfax County

Ray Kroc FAST FOOD ENTREPRENEUR

Raymond A. Kroc built a franchise on the success of the McDonald brothers in San Bernadino, California. Their restaurant allowed customers to park and walk directly to a window to order food. From an airplane, Kroc surveyed suburban Chicago, looking for an appropriate site. In 1955, he opened his first restaurant in Des Plaines, Illinois. Kroc wanted to profit from the growing affluence of the suburbs and the driving that suburban life required.

The connection between eating and transportation infrastructure was not new. Years earlier, Howard Johnson pioneered the chain restaurant as a roadside landmark catering to the highway trade. Jack-in-the-Box, founded in San Diego in 1950, was a drive-through; the entrepreneurs who founded Burger King in 1953 also studied the McDonald brothers' business.

Fast food corporations continue to expand along with the suburbs and develop new opportunities. In 1993, McDonald's opened the first of hundreds of "co-branded" facilities—restaurants attached to gas stations, usually located near highways in rural areas.

Americans are the heaviest people in the industrialized world. Sixty-one percent of the population is overweight. Between 1976 and 1994, the percentage of obese Americans went from 47% to 56%; the percentage of overweight children doubled. A recent report by researchers at the Centers for Disease Control and Prevention blames the suburban lifestyle for the increased occurrence of obesity, diabetes, and asthma.

WHO'S RESPONSIBLE?

Sam Walton CHAIN STORE MOGUL

The postwar "retail revolution" reached a peak in 1962 with the founding of Target, K-Mart, and Wal-Mart. Thanks to increased income and easy credit, shoppers streamed to stores and malls that they could reach one way: by car.

Before Sam Walton began the discount chain that now includes stores in 50 states and partnerships in countries including China, he attempted to develop a suburban shopping mall in Little Rock, Arkansas. Walton understood the essential role of transportation infrastructure, in terms of both his customers and the need to distribute goods to his stores. In the late 1950s, however, roads weren't fast enough for Walton, who flew his own plane between stores.

From the beginning, the strategy of Target and K-Mart was to place stores in the growing suburbs of metropolitan areas. Wal-Mart focused on southwestern and midwestern towns and small cities where there was little competition from national chains—and where stores would be accessible by car from miles around.

In the 1980s, as Wal-Mart expanded, its method was to surround cities with suburban stores. Today, ten Wal-Mart stores ring Washington, D.C., most within a radius of about 30 miles; there are none in the District.

In some areas, Wal-Mart has begun replacing individual stores with gigantic "supercenters." These require most customers to drive longer distances to shop.

The corporations with the highest revenues in the U.S. are Wal-Mart, Exxon Mobil, General Motors, and Ford. They base their businesses on the supremacy of the road in American life.

Joe Alexander COUNTY OFFICIAL

Joe Alexander represented the Lee District on the Fairfax County Board of Supervisors from 1964 until 1996. He developed an expertise in transportation and served on the Metro board for 24 years. Alexander was one of the people behind Lee District planning, which included mass transit, roads, and real estate development. This work was behind the development of the area around the Franconia-Springfield Metro station, now part of the Joseph Alexander Transportation Center.

In the Providence district, officials were more concerned about right-of-way costs and serving existing towns such as Vienna. The Fairfax County Board of Supervisors did not take into account their Planning Commission's prediction that Tysons Corner would develop into a business center; the Metro does not go there. South of Vienna, local residents were able to halt development at their Metro stop despite the social and environmental consequences for their county and region.

Washington's Metro, a heavy-rail subway system that opened in 1976, today has 83 stations and 103 miles of track. In 1970, there were only nine metropolitan areas with mass transit rail systems such as subways, streetcars, and commuter rail. Today, there are 26. The impetus for these systems is the traffic that accompanies suburban expansion. In 1982, the average American spent 16 hours sitting in traffic. In 1997 that number rose to 45.

Most post-war systems are light rail, although it is far from clear that light rail systems work in low-density suburbs designed for driving. New rail lines that are not part of a mass transit network have limited success because they are not easily reached or integrated into communities. Most cities with light rail lines, such as Portland, San Diego, and St. Louis, report relatively low use. Still, there are signs that the traditional automobile suburb may be changing. In 2000, for the first time in decades, the percentage of mass transit use increased more than the automobile use.

Suburb-to-suburb transportation has become a major issue. About half the population of Fairfax County now works within the county; only 19% commute to Washington, D.C. This is where, however, the road systems and Metro were designed to bring people. In late 2001, Loudoun County supervisors met with their counterparts in adjoining Prince William County—"for the first time in memory," according to the *Washington Post*—in order to discuss county-to-county transportation.

Evening rush hour, Vienna Station, Washington, D.C. Metro

Daniel Patrick Moynihan U.S. SENATOR

Daniel Patrick Moynihan, New York Senator from 1977 to 2001, was chief author of the Intermodal Surface Transportation Efficiency Act of 1991, or ISTEA, renewed in 1999 as TEA-21. With this piece of federal legislation, Moynihan helped to change road design throughout the nation.

TEA-21 requires planners, including civil engineers, to consider people, roads, and environment together. In other words, it redefines "transportation" as something more than highway construction. The term now refers to the building and maintaining of infrastructure that reflects multiple users and issues: pedestrians, cyclists, cars, energy conservation, pollution, and cultural landscapes. TEA-21 gives the states greater flexibility and requires public involvement in transportation planning.

ISTEA and TEA-21 have had an impact in the Washington, D.C. suburbs. Between 1994 and 1996, the public, encouraged by meetings and a web site, participated in a planning study of the Dulles Corridor. Existing roads are not capable of handling predicted traffic increases; rather than continue to build roads, the study recommended a rail system. In 2000–2002, a second public participation program studied this option.

Public participation in Dulles Corridor planning

Parris N. Glendening
STATE GOVERNOR

According to Parris N. Glendening, Governor of Maryland since 1985, "Sprawl is a disease that is eating away at the fabric of our communities, creating a hidden debt of unfunded infrastructure and services, social dysfunctions, urban decay and environmental degradation." He is positioning Maryland as a national leader in "smart growth," which is intended to encourage mixed land use and development in existing communities, slow the consumption of land, and provide the public with transportation choices that include walking.

"A housing development plopped down in the middle of a farm preservation area becomes a cancer that threatens the surrounding region."

—Maryland Governor PARRIS N. GLENDENING, 2001

Larry Cloyed TRANSPORTATION ENGINEER

Larry Cloyed's official title, Assistant Resident Engineer, only hints at the complexity of his job as senior manager of the Springfield Interchange project. While directly managing about 60 people, Cloyed enables municipal agencies, public officials, and his supervisors at the Virginia Department of Transportation to all work together.

Cloyed was born in Washington, D.C. and grew up in Centreville, Virginia, then a rural part of Fairfax County. He began his engineering career as a land surveyor and worked in positions of increasing responsibility for the Forest Service, the Federal Highway Administration, and for the last 14 years, the Virginia Department of Transportation.

Cloyed's top challenge is to manage the efforts of a network of people so that roadways remain safe during the multiple phases of Springfield Interchange construction. He must ensure that motorists can drive through with a minimum of delay. And he must monitor modifications that take place during construction, when the results of years of planning meet the real world. In a new and increasingly important role for civil engineers, Cloyed is in charge of public outreach. He oversees an information center in the Springfield Mall. It attracts several thousand visitors a month.

The Springfield Interchange, which links I-395, I-95, and I-495, will include 41 miles of roadway and take 8 years to build.

SPRINGFIELD INTERCHANGE

The car-centered lifestyle will soon have a new monument: the Springfield Interchange. When finished in 2006, this road network, which is really three interchanges, one on top of the other, will have 50 new bridges. Some sections of the highway will be 24 lanes wide. At a cost of an estimated $600 million, the project will allow drivers to move their cars from one road to another. Because both Interstate and local traffic—up to 400,000 cars a day—speeds through, changing lanes has become a complicated and dangerous operation.

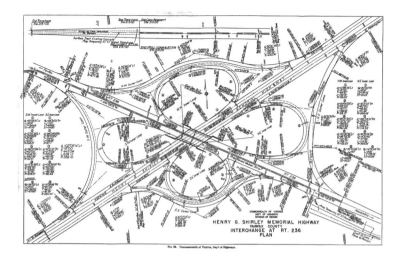

TRAFFIC JAM

Do new roads cure traffic jams or cause them? Transportation engineers and public officials have long debated this question. What is certain is that traffic congestion became a fact of life beginning in the mid-1920s. At first, experts were confident they could improve traffic with added lanes or better management. Now, they aren't so sure. Metropolitan areas that build the most highways have the worst traffic, according to several studies. This suggests that highways cause, and rarely cure, traffic.

How could this be? New highways encourage more driving. Homes and businesses appear on the land made accessible by new roads. This leads to even more driving. The latest thinking is that traffic congestion can best be solved by "congestion pricing" – charging fees for driving at rush hour.

In the suburbs, roads and highways make up a large percentage of public land— territory dominated by private vehicles. Because roads now provide public access to almost everywhere people want to go, engineers design infrastructure—including water lines, sewer, storm water, electrical, telephone, and cable lines, alongside or under roads.

A complex and ever-changing group of laws govern these systems. Some, like water supply and sewerage, are completely public. Municipal governments own the water lines and their workers install them. Regulated private companies, such as Verizon and Dominion Virginia Power, own communication and power lines, but must cooperate with public authorities, usually the highway departments, to install and maintain them. In the past decade, governments have encouraged competition among companies to supply services; this experiment is called "deregulation," although it is heavily regulated.

> "We want to avoid in ten years the Silicon Valley experience."

—Fairfax County Executive JOHN F. HERRITY, speaking about transportation, affordable housing, and education, 1985

Top: Interchange of Henry G. Shirley Memorial Highway, Virginia's first Interstate, 1949
Bottom: Springfield Interchange under construction

Bill von Meister
WEB ENTREPRENEUR

America Online chief Steve Case noted that William Von Meister was vital to the creation of web-based commerce. Von Meister was a "compulsive entrepreneur," according to the *Washington Post*. After growing up in New Jersey, and attending Georgetown and American Universities, von Meister stayed in northern Virginia, where he was one of a network of technologists, investors, and former government employees and consultants. In the Washington, D.C. suburbs they built an e-commerce capital.

In Manhattan, next to Central Park, a new skyscraper will contain the corporate headquarters of the new conglomerate AOL Time Warner. New York's status as an international media capital is partly behind this unusual move from suburb to city. Corporate chairman Steve Case has said that he will keep his home and office in Northern Virginia. The World Wide Web and services provided by companies such as AOL enable the ongoing movement of business away from cities and into the suburbs.

AOL headquarters, old and new buildings, Loudoun County

Von Meister's ventures included the first online consumer information service, The Source (1978), as well as networks able to send music and video games to people's homes via satellite and cable. In 1983, von Meister met Case at a trade show in Las Vegas and hired him because of his marketing experience. Through this opportunity, Case became one of the people behind the firm that became America Online. Based in Fairfax County, it moved in 1996 from Vienna to a Loudon County office building that became the starting point of an expanding AOL office park. AOL properties also include a new computer center in Prince William County.

WHO'S RESPONSIBLE? 31

New Homeowners

Why move to the farthest reaches of suburban Washington, D.C.? This couple likes the rural nature of the area with its mountains views. He previously lived on a farm and works at home; she works in the District of Columbia several days a week. While house hunting, they discovered that each 20 minute decrease in driving time to Washington meant a $100,000 increase in the price of a comparable property.

New subdivision, Loudoun County

BUILDING A NEW HOME

Each new home must be linked to the infrastructure before its owners can move in. Fairfax County requires the following permits.

A building permit is required for construction of architectural and structural elements of new homes.

An electrical permit is required for all electrical installations.

A mechanical permit is required for installation of all elements and appliances associated with heating and air conditioning systems.

A plumbing permit is required for installation of all elements and appliances associated with plumbing and gas piping systems.

A Virginia Department of Transportation permit is required if the lot is located on a public road with the driveway entering a state road.

A sewage disposal system construction permit is required from the Health Department if the single family dwelling is constructed with an individual sewage disposal system.

A water supply construction permit is required from the Health Department if the single family dwelling is constructed with an individual well.

Farms, like cities and suburbs, are built environments. The economic and land-use policies behind the rise of vast agricultural complexes has sped the decline of smaller tradional farms, whose land has become the foundation of the new suburbs.

Farm and residential subdivision, Loudoun County

RESPONSIBILITY

Road building—like the construction of all infrastructure—is a cultural, political, and technological activity. For more than a century, road supporters have debated the options: whether to focus on rural roads for farmers, urban roads in cities, secondary roads around towns, primary roads between cities, or the limited-access highways known as Interstates. Should general taxes or tolls pay for roads? How important is walking and bicycling? What should roads look like? What are the most appropriate materials? Is curved or straight better? Should cars or trucks have priority? Who builds roads for corporations that want headquarters in the suburbs? Should roads be built in densely populated or isolated areas? Is there a solution to traffic jams? Should population, income, traffic, or land area determine the distribution of funding for roads?

These are a few of the questions that determine the shape of the United States. There is no single correct answer. Winners of past road debates are responsible for the quality of life in America and will be for a long time to come.

Atlanta, Georgia

A walk across the street seems natural, but it is an engineered activity. Paving, traffic light, crosswalk, warning sign, lighting, and perhaps, sidewalk: these make up the infrastructure of the pedestrian experience.

Is It Safe?

Civil engineers conduct safety studies, analyze traffic patterns, and seek to understand and influence the behaviors and the environments that cause crashes. Along with pedestrians, drivers, police, insurers, and government officials, engineers make decisions that determines what safety is and who benefits from it. They strive to satisfy the conflicting needs of pedestrians and drivers.

Pedestrians conduct a complex risk assessment every time they enter the roadway. "Is it safe?" is a private question with public consequences.

Background: Mid-block crossing, Atlanta

Safety Today: Atlanta	38
History of Safety: New York City	50
Future Safety: Portland, Oregon	58

Crossing

Almost everybody does it, some several times a day. It seems simple but it's not.

If crossing the street was a simple activity, why were 4,739 people killed and 78,000 injured while crossing U.S. streets in 2000?

Getting from one side to the other is dangerous because roads were designed for cars, not people. Growing public recognition of the social and environmental impacts of car-centered design is shifting the focus to feet and bicycle wheels.

> "Society appears to react more strongly to infrequent large losses of life than to frequent small losses."
>
> —Economist PAUL SLOVIC, "Facts and Fears: Understanding Perceived Risk," 1980

WESTON THE PEDESTRIAN.

Around 1900, spooked horses killed hundreds of Americans each year. The addition of automobiles to horse and trolley traffic caused a dramatic increase in the pedestrian death rate.

Atlanta, Georgia SAFETY TODAY

"A pedestrian in Georgia is more than twice as likely to be killed by a stranger with a car than a stranger with a gun."

—PEDS (Pedestrians Educating Drivers on Safety), Atlanta

"When an accident occurs, it is a breakdown in the system."

—Engineer and psychologist DAVID SHINAR, *Psychology on the Road*, 1978

Accident

Some say there is no such thing as an accident. Traffic engineers now use the term *crash* to acknowledge that chance is not an adequate explanation for death or injury. Instead, they ask, "why did failure occur?" Examination of pedestrian infrastructure provides the answers.

Above: Sample traffic collision report, 1982

Facing page: Suburban intersection, Atlanta

Suburban Atlanta

Atlanta has been dubbed "sprawl city" by sociologist Robert Bullard. Its population increased from half a million in 1940 to over three million today. Most Atlantans live in suburbs whose expansion is unrestrained by natural barriers. Atlanta is the least densely populated metropolitan area in the United States.

During the 1990s, Atlanta's physical boundaries doubled. It emerged as the business and financial center of the southeast, as well as a transportation and communications hub that is home to CNN and the nation's busiest airport.

In order to manage its growing transportation needs, Atlanta created a regional highway plan in 1946—six years before passage of the Federal Interstate Highway Act. In recent years, unlimited suburb growth, without the transportation planning that would support such growth, has increased dependence on private cars as the primary means of transportation. The Atlanta region has become less friendly to non-automotive modes of travel, such as walking. With that comes an increase in the number of pedestrian deaths.

Pedestrian Atlanta

The Centers for Disease Control and Prevention recently identified the Atlanta metropolitan area as the nation's third most dangerous for pedestrians, after Ft. Lauderdale and Miami. Between 1994 and 1998, 309 pedestrians were killed in the Atlanta area—an increase of 13% compared to a national decrease of 9.6%.

Crosswalks, Atlanta

THE PEDESTRIAN Walking and Flying

The Surface Transportation Policy Project, a group that promotes a people-oriented rather than vehicle-oriented focus in transportation policy, calculates that walking is far more dangerous than driving or flying per mile traveled.

Fatality Rate per 100 Million Miles Traveled: **DRIVING 1.4 FLYING 0.2 WALKING 49.9**

For each mile traveled, walking is 36 times more dangerous than driving, and over 300 times more dangerous than flying.

Safety is a relative concept. There is no specific number of fatalities that defines *safe*. Or is there?

Driving More, Walking Less

Driving has become safer for drivers and their passengers. The total number of deaths, and deaths per vehicle mile traveled, have declined because of safer car design and safer driver behavior—increased use of seat belts and child safety seats, along with a decreasing number of drunk drivers. Total traffic deaths have decreased even though the number of cars continues to increase.

Meanwhile, the risk to street-crossing Americans may have increased. The U.S. Department of Transportation notes that the decline in the number of pedestrian deaths is probably due to the fact that people walk less.

Roads and sidewalks, Atlanta

Standard pedestrian speed is 4 feet per second.

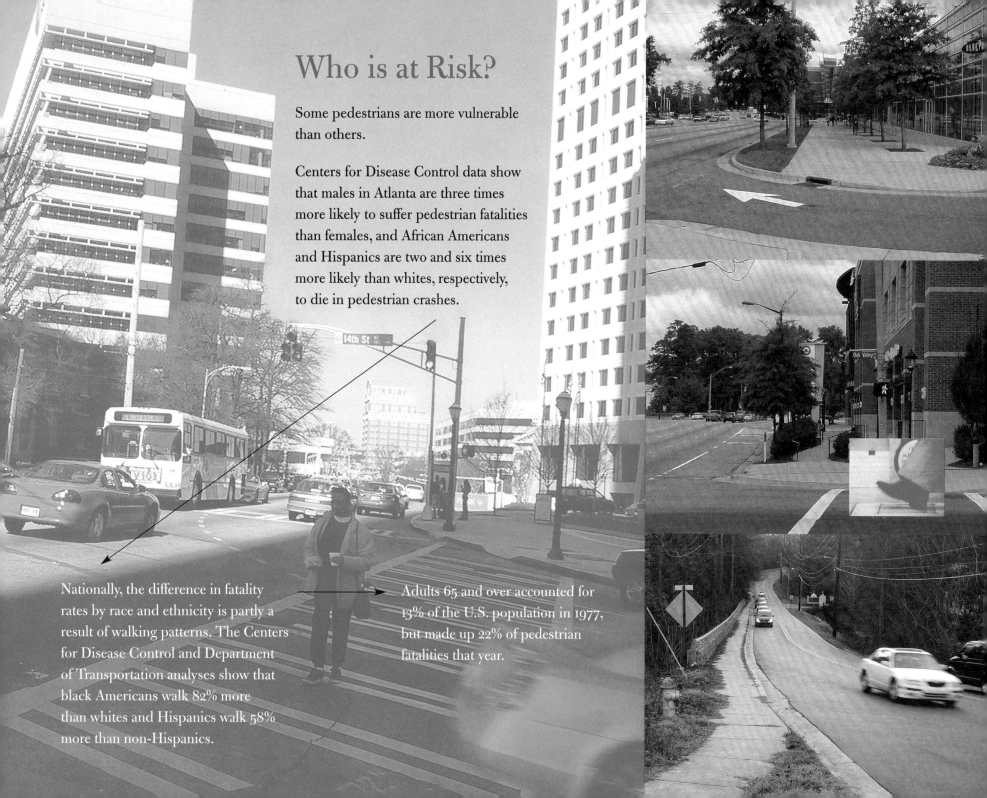

Who is at Risk?

Some pedestrians are more vulnerable than others.

Centers for Disease Control data show that males in Atlanta are three times more likely to suffer pedestrian fatalities than females, and African Americans and Hispanics are two and six times more likely than whites, respectively, to die in pedestrian crashes.

Nationally, the difference in fatality rates by race and ethnicity is partly a result of walking patterns. The Centers for Disease Control and Department of Transportation analyses show that black Americans walk 82% more than whites and Hispanics walk 58% more than non-Hispanics.

Adults 65 and over accounted for 13% of the U.S. population in 1977, but made up 22% of pedestrian fatalities that year.

Toddling

In 1999, 567 pedestrians age 15 and younger were killed in the United States, including 163 children under four years old. Motor vehicles are the leading cause of death for children.

Adults often blame the child who has been hit by a car. Children are faulted because they "dart out" from between parked cars. Blaming the victim, however, leads to prevention methods that focus on the child, rather than structural improvements or driver behavior. Meanwhile, the efficacy of education programs aimed at children has not been proven.

The lack of a safe walking environment for children, along with the increasing dependence on cars, has decreased the time that children spend on their own, walking to school. Researchers in England report that while 80% of seven and eight year olds went to and from school on their own in the early 1970s, fewer than one in ten were doing so in the 1990s.

Atlanta crosswalks

Jaywalking

This term comes from the word *jay*, which once referred to a country cousin who was unfamiliar with city ways. Now, the term refers to a pedestrian who crosses without regard for rules.

Who is at fault? The idea that a jaywalker hit by a car might be at fault sparks controversy. Whose behavior needs modification: drivers or pedestrians?

Walk/Don't Walk

In one study, only 2.5% of a sample of 400 pedestrians understood the meanings of the flashing and steady WALK/DON'T WALK signals. Many believed that turning cars would not cross their path.

Left: Traffic engineering form

Right: Traffic engineers added yellow lights to traffic signals in the 1950s.

Pedestrian Speed

Engineering guidelines for the timing of traffic lights at pedestrian crossings assert that pedestrians cross at four feet per second. Older pedestrians, however, or individuals with disabilities, or those carrying children or shopping bags, may need longer to cross. Bad weather slows crossing time, too.

It is estimated that one-third of pedestrians walk more slowly than the 4-feet-per-second design standard.

Measuring Safety

It is difficult to compare the safety of street crossing to other modes of travel because there is a limited amount of data on who walks, for how long, and where. The first attempt to quantify the practice of walking in the United States was undertaken in 1995, when the federal government commissioned the National Personal Transportation Survey. This signaled a shift in focus from driver to walker.

The U.S. Department of Transportation reports that the 4,906 pedestrians killed in 1999 represent 12% of the 41,611 people killed in traffic crashes that year. On average, a pedestrian is killed by a motor vehicle every 107 minutes and injured every six minutes.

The Idea of Safety

Safety is often viewed as a limited resource gained through trade-offs among time, money, and convenience. "Is it safe?" may be more accurately phrased, "How much safety and for whom?"

Safety is usually measured solely in terms of crashes, deaths, and injuries. Public perception of safety, however includes personal experience, knowledge, and myth.

Left: "Pavement reminder" for children, 1940s

Above: Safety warning, 1940: a painted cross indicated the site of a pedestrian fatality

Right: "Safety" is a measure of death and injuries, not safe crossings

Facing page: Insurance company advice, 1940

"So dangerous it's safe"

Many pedestrians are familiar with the experience of almost getting hit by a car. These "near-misses" don't show up in pedestrian fatality or injury statistics, although they are crucial for understanding safety.

When an intersection is so risky that pedestrians avoid it, or cross with extreme caution, traffic engineers say that "it's so dangerous it's safe." A street's low number of pedestrian crashes may indicate only that few people attempt to cross it, not that it is safe.

PEDESTRIANS

A DECALOGUE OF SAFETY
for Pedestrians

1. Cross Only at Crosswalks

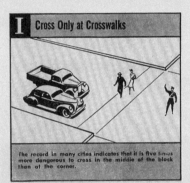

The record in many cities indicates that it is five times more dangerous to cross in the middle of the block than at the corner.

2. Wait on the Sidewalk

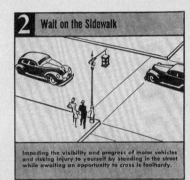

Impeding the visibility and progress of motor vehicles and risking injury to yourself by standing in the street while awaiting an opportunity to cross is foolhardy.

3. Cross on the Proper Signal

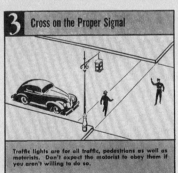

Traffic lights are for all traffic, pedestrians as well as motorists. Don't expect the motorist to obey them if you aren't willing to do so.

4. Be Sure the Way Is Clear

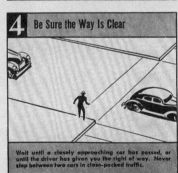

Wait until a closely approaching car has passed, or until the driver has given you the right of way. Never step between two cars in close-packed traffic.

5. Be Doubly Alert During the First Few Steps

Seventy-five per cent of pedestrians in accidents are hit before reaching the middle of the roadway, with absent-mindedness the greatest single cause.

6. Cross Within the Crosswalk

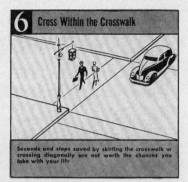

Seconds and steps saved by skirting the crosswalk or crossing diagonally are not worth the chances you take with your life.

7. Walk to the Right in Crosswalks

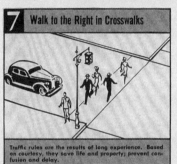

Traffic rules are the results of long experience. Based on courtesy, they save life and property; prevent confusion and delay.

8. Walk, Don't Run

Needless hurry afoot is often as dangerous as needless speed in a car. Don't start across unless you are sure you can make it safely at a walk.

9. On Rural Roads, Walk Facing Traffic

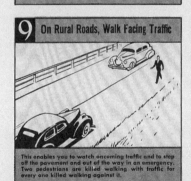

This enables you to watch oncoming traffic and to step off the pavement and out of the way in an emergency. Two pedestrians are killed walking with traffic for every one killed walking against it.

10. After Dark, Wear Something White

You protect yourself by making certain you will be seen, especially when walking along rural highways at night. Wear a white shirt, shoes, or dress—or carry a light, an open handkerchief, or even a newspaper.

IS IT SAFE? 45

THE ROAD Suburban Streets

Behind the image of suburbia as a peaceful, safe place lies an uncomfortable reality: suburban roads and streets are dangerous for pedestrians.

Suburban areas have a higher rate of pedestrian fatalities than heavily traveled urban streets, because in the 1950s and 1960s, developers and civil engineers designed streets in the suburbs based on a car-centered highway model. They did not design streets for walking and public policy did not demand it.

Higher speeds on suburban arteries and collector streets, and drivers' lack of attention to pedestrians add to the higher fatality rate.

"WHAT'S BECOME OF YOUR BROTHER BILL, ED?"
"HAVEN'T SEEN HIM FOR TEN YEARS. HE LIVES ACROSS THE STREET."
1925

Mid-Block Crossing

Most pedestrian crashes occur at intersections, but the lower speed of traffic makes them less deadly. There is a higher rate of pedestrian fatalities at mid-block because that is where cars travel at higher speeds. Fewer pedestrians cross at mid-block, but those that do are at greater risk of being killed.

Pedestrians cross at mid-block because crossing opportunities are spaced far apart. They trade safety for time and convenience.

Facing page: Suburban road, Atlanta

Right: Advice for nighttime walking

Sidewalk

A sidewalk is an essential component of the infrastructure of pedestrian safety.

Residential areas without sidewalks have only 2.7% of the pedestrian traffic, but account for 23.4% of the pedestrian collisions, according to a report by the Institute of Transportation Engineers.

Developers build suburbs without sidewalks in order to save money. Some assume that sidewalks are not needed because everyone drives. A lack of sidewalks may be intended to discourage people who don't own cars from entering a neighborhood.

Visibility

Most pedestrian and bicyclist fatalities occur between six and nine in the evening, when visibility is low and there are more people and cars on the streets. One response of safety experts has been to suggest that pedestrians wear light-colored or reflective materials at night.

THE DRIVER

"Every two miles, the average driver makes four hundred observations, forty decisions, and one mistake."

—Journalist MALCOLM GLADWELL, "Wrong Turn," 2001

"The accident- and violation-free driver is more mature, conservative, and intellectual in his interests and tastes, has a higher aspiration level, and is the product of a happier family background... To date, neither personnel nor techniques of psychotherapy [are available] to correct the underlying causes of accident likeliness."

—Engineers CLARKSON OGLESBY AND LAURENCE HEWES, *Highway Engineering*, 1963

Home

In studies of driver behavior, psychologist and engineer David Shinar links road congestion and safety. Traffic jams lead to frustration and aggressive driving. The more time drivers spend in traffic, the more they treat the car as an extension of the workplace and the home. People at the wheel dangerously distract themselves from driving: they talk on the telephone, drink, eat, write, and even read.

Although roads are the most public of spaces, the fact that automobiles are private may encourage drivers to engage in home activities. They endanger the lives of pedestrians and bicyclists because they feel they own the road, as they own their cars.

Speed Kills

A little speed is enough to kill a pedestrian.

Pedestrians hit by cars going over 35 mph are usually killed, or sustain life-threatening injuries. Pedestrians hit by cars going less than 20 mph are usually not injured seriously.

Background: "Goodness, the vehicles of Mrs. Striker and Mrs. Bangs have entered into a collision." *Saturday Evening Post* cover, 1956

Speeding, Seeing, and Stopping

Speed affects the driver's ability to halt the car without killing the pedestrian. The stopping distance of a car going 25 mph compared to a car going 40 mph is the difference between a fatal crash and a "near-miss."

Estimating the speed of an approaching car

There are several reasons for this. A driver can spot a pedestrian only a limited distance ahead; after seeing a person on the road, he or she requires at least half a second to apply the brake. The car requires additional time to stop. At 25 mph, a driver has time to avoid hitting a pedestrian. At higher speeds, there is not enough time for seeing, braking, and stopping to take place.

Bad weather conditions, curves in the roadway, hills, or obstructions can shorten sight distances and make stopping distances even longer.

"Almost one of every three traffic fatalities is related to speeding, and speeding is a safety concern on all roads, regardless of their speed limits . . . Almost 50 percent of speeding-related fatalities occur on lower speed collector and local roads."

—FEDERAL HIGHWAY ADMINISTRATION, *Speeding Counts on All Roads*, 2000

Under the Influence

The dangers of drinking and driving have been well established and publicized. This has led to changes in legislation and driver behavior.

Intoxicated drivers are especially dangerous to pedestrians. Pedestrian fatalities are more likely to be connected to alcohol than are fatalities involving drivers and their passengers, according to the National Highway and Transportation Safety Administration.

Skid mark width indicated by pen in crash investigation photograph

In 1998, according to the Department of Transportation, one-third of pedestrian victims 14 or older had blood alcohol content high enough to be considered intoxicated. But researchers have not determined if walking while intoxicated is a primary cause of pedestrian crashes.

Nearly half of pedestrian fatalities involve intoxication of driver, pedestrian, or both. In about ten percent of fatal pedestrian crashes, the driver is intoxicated.

THE CONSTRUCTION OF SAFETY
New York, New York

Background: 74th Street and Central Park West, New York: site of the first recorded pedestrian death caused by an automobile driver, 1899

Left and facing page: The plight of pedestrians—already a magazine feature in 1925

How safe is safe enough?

The appearance of the automobile at the end of the 19th century brought a new, violent danger to the lives of New Yorkers. At various times during the last century, thousands of deaths led the public, civil engineers, police, government officials, and insurers to recognize pedestrian risks. They devised ways to spare the lives of New Yorkers trying to get from one side of the street to the other.

"We speak of traffic flow, using a metaphor to rivers. But nowhere in nature do two rivers cross. Basically, it is an unnatural situation that requires tremendous human ingenuity to control."

—Engineer DAVID GURIN, New York City Department of Transportation Bureau of Planning and Research, 1981

Cars vs. Pedestrians

In the United States, the first crash resulting in a pedestrian death occurred on September 13th, 1899, at Central Park West and 74th Street in New York City. A car struck and fatally injured Henry H. Bliss as he was getting off the Eighth Avenue trolley. Bliss, 68, was a Vermont native who had been a real estate dealer in New York for 35 years. The *New York Times* reported that the corner where the accident happened was known to the trolley line motormen as "Dangerous Stretch."

This was the beginning of a 100-year war between drivers and pedestrians in the streets of New York.

New York has the highest number of pedestrians killed in traffic accidents of any city in the United States. The annual death toll has ranged from a high of over 500 in the early 1970s to about 200 in the late 1990s. In a city with a population of eight million, that's a relatively low rate, although nobody is likely to call the streets of New York safe for walkers.

"I'LL BETCHA THOSE WOODS ARE FULL OF PEDESTRIANS!"

IS IT SAFE? 51

Class Crash

At the beginning of the 20th century, only the rich could own cars. When wealthy New York drivers encountered New York pedestrians, clashes resulted. In December 1899, the *New York Times* reported on some of the New Yorkers who had recently obtained licenses to operate their automobiles: Alfred G. Vanderbilt, Harry Payne Whitney, and A.R. Shattuck, the son-in-law of a former mayor.

Manhattan street grid: Commissioner's Plan, 1807

New York Grid

The New York Plan of 1811, which established the street grid system, was intended to speed land development, avoid property-line confusion, and allow for circulation of air, which was believed to improve public health. The interests of businessmen and landowners were paramount in the establishment of the grid. Then, as now, public investment in infrastructure formed the foundation of private wealth.

Consisting of twelve avenues and 155 streets, the New York grid set the pattern for what is today one of the greatest pedestrian cities in the world. The grid was in place for a century when the automobile began to dominate the streets.

Before the car raced into New York life, streets were crowded and difficult to cross, but not as dangerous as they are today.

Death in the Street

Under the headline, "Boy Crushed to Death By a Motor Vehicle," a 1902 front-page story in the *New York Times* reported that "a heavy motor vehicle containing a party of four women enjoying an afternoon ride yesterday dashed around the corner of Fortieth Street and Tenth Avenue so close to the curb that seven-year-old Joseph Buscher, son of Cord Buscher, a butter and milk dealer at 530 Tenth Avenue, who was stooping in the gutter reaching for a stray marble, was thrown down and crushed to death in less than a second." A witness said that "it seemed to me that it passed almost in a flash. I don't see how any one could have got out of the way whether they were looking or not."

Traffic Engineers

In the early part of the 20th century, the New York City Police Department had jurisdiction over traffic regulations. In 1924, the Police Commissioner issued regulations to improve traffic safety entitled "Thou Shalt Not Kill." Speed limits for passenger vehicles were set at under 20 mph in the city and less than eight mph in congested intersections.

Plan for traffic tower, New York, 1922

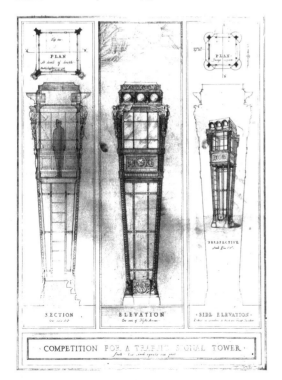

Traffic Light

An early innovation in traffic safety was the traffic signal. Seven bronze traffic towers were placed along Fifth Avenue in 1922 and remained there through the remainder of the decade.

In 1931, the more familiar traffic lights were installed on Fifth Avenue. These early signals were topped by a bronze figure of Mercury, the god of travel. Many pedestrians probably wished they had Mercury's winged feet.

Traffic light with figure of Mercury, Fifth Avenue, New York, 1930s

Pedestrian Signals

In 1952 and 1953, New York City won the AAA National Pedestrian Protection Contest, as the safest large city for pedestrians. It had a rate of 5.5 pedestrians killed per 100,000 people, compared to 7.1 deaths in comparable urban areas.

In 1955, the first traffic signals designed specifically for pedestrians were installed in New York City. Today, pedestrian signals use pictures instead of words.

Installation of pedestrian signal, New York, 1950s

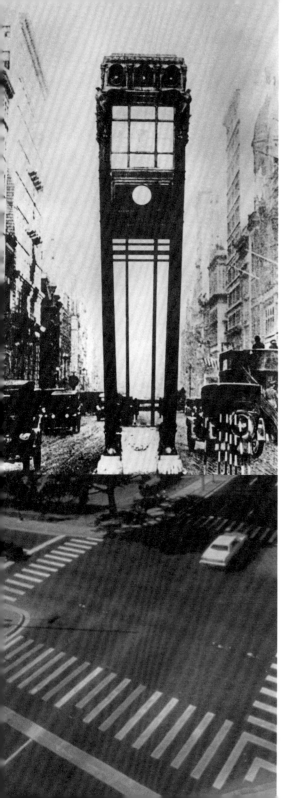

Pedestrian Engineers

Henry Barnes was Traffic Commissioner in New York from 1961 to 1968. Barnes brought many safety innovations to New York, including the involvement of the media and businesses in the promotion of pedestrian safety. He is noted for the "Barnes Dance," a red signal interval for all cars at an intersection that allowed pedestrians to cross in all directions, including diagonally.

The work of traffic engineer John Fruin in New York in the early 1970s led to changes in thinking about how street design might accommodate pedestrians. His tactic was to apply traffic flow concepts to people. Planners and engineers began to integrate pedestrians into their designs. Increased safety was not the main concern, but it was an unintended benefit.

Above left: Fifth Avenue traffic tower, New York, 1920s

Left: Intersection with crosswalks

Below: Diagram showing moderate level of pedestrian crowding

Above right: New York Traffic Commissioner Lloyd B. Reid at installation of first reflectorized stop sign, Brooklyn, 1951

Below right: Painting crosswalks, New York, 1950s

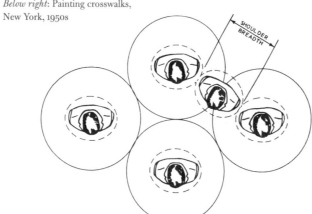

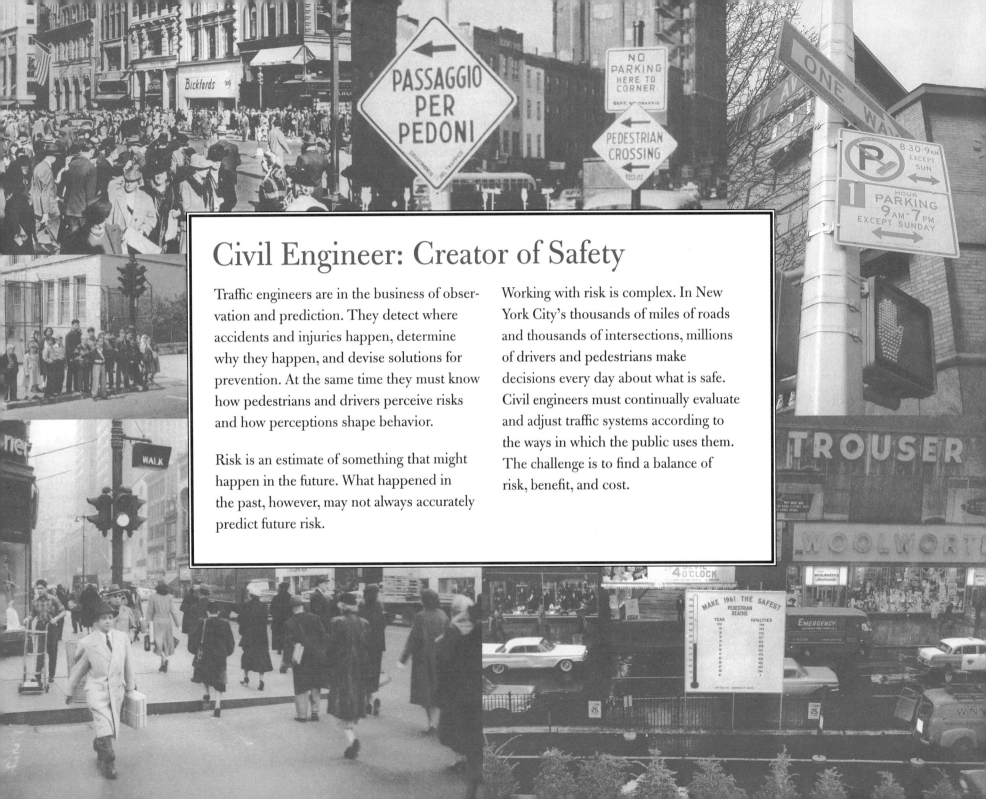

Civil Engineer: Creator of Safety

Traffic engineers are in the business of observation and prediction. They detect where accidents and injuries happen, determine why they happen, and devise solutions for prevention. At the same time they must know how pedestrians and drivers perceive risks and how perceptions shape behavior.

Risk is an estimate of something that might happen in the future. What happened in the past, however, may not always accurately predict future risk.

Working with risk is complex. In New York City's thousands of miles of roads and thousands of intersections, millions of drivers and pedestrians make decisions every day about what is safe. Civil engineers must continually evaluate and adjust traffic systems according to the ways in which the public uses them. The challenge is to find a balance of risk, benefit, and cost.

Crossing the Boulevard

Queens Boulevard, once a small, lightly traveled road, is now one of the widest in New York City. With eight lanes of traffic moving at highway speeds, the boulevard has become notoriously dangerous for pedestrians. Since 1993, 73 pedestrians have been killed crossing it.

Queens Courier, 2001

A study of boulevard safety by Allan B. Jacobs and colleagues at the UC-Berkeley Institute of Urban and Regional Development found that boulevards with wide traffic lanes and long blocks are associated with higher speeds and more mid-block crossings by pedestrians.

The installation of warning signs, pedestrian barriers, longer pedestrian crossing times, and the enforcement of traffic laws may help build the perception that Queens Boulevard is safe.

Risk and Control

Most people have a fairly accurate idea of the dangers that await the pedestrian. Each person's feelings about risk, however, are the key to determining if, when, and how he or she will cross a street.

Everyone weighs risks and benefits before engaging in everyday activities. A study by risk analyst Paul Slovic and his colleagues found that dread, control, and the potential for catastrophe affect the average person's risk perception. People are more averse to a risk they feel they cannot control.

Slovic concluded that evidence about real risk may not be enough to change public perceptions and behaviors, because what is most important is whether or not people feel in control.

"There is no single all-purpose number that expresses 'acceptable risk' for a society…there are no value-free processes for choosing between risky alternatives."

—Decision science expert BARUCH FISCHOFF

Pedestrian safety education, New York, 1950s

Pedestrian and cyclist at 74th Street and Central Park West, New York

56 ME, MYSELF AND INFRASTRUCTURE

The Safety Industry

In the early part of the century, the insurance industry became involved in traffic safety research and education. They recognized that reducing the number of traffic deaths was the right thing to do; the first compulsory driver insurance law went into effect in Massachusetts in 1927. Since then, most states have adopted a requirement that motorists carry insurance coverage. Some critics state that no-fault insurance protects the at-fault motorist against enormous liability and may have significant impact on driver behavior in relation to pedestrians.

The auto industry became involved in safety and community service activities long ago. The American Automobile Association Foundation for Traffic Safety, established in 1947, included traffic engineers motivated by a desire to improve safety, save lives, and maintain a positive image for the automobile.

AAA pedestrian safety posters, 1955

Record album, c. 1967

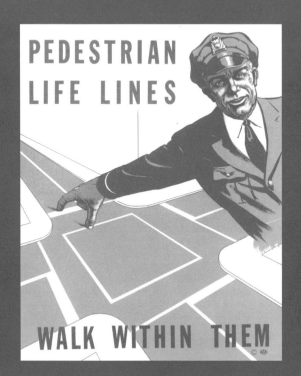

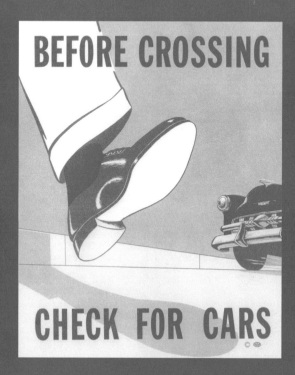

IS IT SAFE? 57

FUTURE SAFETY

Engineering Safety

The community—the public, engineers, government—together decide what's safe and what's not, by balancing competing interests, costs, and benefits. It's an ongoing negotiation.

Portland, Oregon, is a community that has been able to reach consensus on methods of maximizing pedestrian safety.

Portland, Oregon

The city of Portland has a history of planning for walkability, beginning with its 1870 street grid design of small 200-foot blocks. Today, Portland is a city with a pedestrian-oriented transportation system facilitated by the state's mandate to coordinate land use and transportation planning, and encouraged by Portland's history of community involvement.

In 1994, Portland adopted a 50-year regional growth and development plan calling for "the development of a true multimodal transportation system which serves land use patterns, densities, and community designs that allow for and enhance transit, bike, pedestrian travel and freight movement."

Third Street, Portland, Oregon.

Background: Textured pedestrian curb cut, Portland

Mt. Hood Freeway

Robert Moses, the master planner of New York City's highways and bridges, proposed the Mount Hood Freeway to Portland. His plan included a six-mile, eight-lane freeway that would have divided the city and resulted in the demolition of 1,750 homes. Community activists, bolstered by evidence showing that the freeway would not relieve traffic congestion, succeeded in blocking the plan in 1974. They gained support by devising a plan to transfer federal highway funding to other local projects.

The history of the Mount Hood Freeway represents the power of communities to say "no" to a car-centered culture.

CASE STUDY: Woodstock Boulevard Pedestrian Plan

The Portland Department of Transportation and its civil engineers undertook the redesign of a major thoroughfare, Woodstock Boulevard, in order to improve the safety and efficiency of the road.

They established a process that allowed people who live and work near Woodstock Boulevard to reach an agreement on its redesign.

This plan to improve safety and quality of life demonstrates that, despite the fact that infrastructure seems inevitable, there are always alternatives available. In a partnership, civil engineers can innovate and advocate change and the public can demand it.

Median island and crosswalk, Portland

Woodstock Boulevard at 44th Street before and after traffic calming

Listening to the Public

The Woodstock Blvd. Pedestrian Plan Community Advisory Committee met over the course of several months to establish the main concerns of residents and help develop the new design.

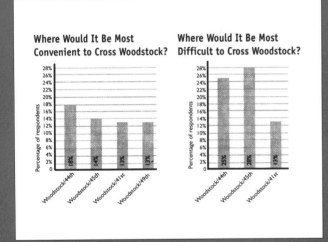

Questionnaire, Woodstock Boulevard Pedestrian Traffic Calming Project

WOODSTOCK BOULEVARD PEDESTRIAN PROJECT

The City of Portland Pedestrian Transportation Program is starting a project along Woodstock Boulevard from 39th Avenue to 52nd Avenue. The project will identify and construct improvements to increase safety and access for pedestrians.

This survey is one of the first steps in obtaining the community's view on pedestrian crossing and safety issues along Woodstock Boulevard.

Your input is very important to ensure the improvements selected and constructed fit the desires of the community. Please take a few moments to fill out this survey.

Your comments should be received by April 25, 1997, for them to be incorporated into the project. Send the completed survey to the Pedestrian Transportation Program using the enclosed postage-paid envelope.

This project is funded through a grant received by the City of Portland Office of Transportation.

If you have any questions about the project, call Chris Armes, Project Manager, at 823-7051/TDD 823-6868.

TELL US ABOUT YOURSELF   PERCENTAGE OF RESPONDENTS

1. **How close to Woodstock Boulevard do you live?** (circle one)

   under 4 blocks 48%   4 - 8 blocks 33%   over 8 blocks 19%

2. **How do you usually get to Woodstock Boulevard?** (circle one)

   Walk 46%   Drive 49%   Bike 4%   Bus 1%

3. **How often do you shop at the businesses along Woodstock Boulevard?** (circle one)

   Daily 51%   Once a Week 47%   Once a Month 2%   Never 0%

4. **How often do you walk along Woodstock Boulevard?** (circle one)

   Daily 35%   Weekly 42%   Monthly 16%   Never 7%

PEDESTRIAN TRANSPORTATION PROGRAM   CHARLIE HALES, COMMISSIONER

Zigzag

Diamond

Shark's Tooth

Arrow

Chevron

Danish Checkerboard

Zebra

Transverse

Traffic Calming

Traffic calming describes methods of balancing the space requirements and convenience of motor vehicles with the safety of pedestrians and bicyclists. Its purpose is to slow traffic and even discourage drivers from using specific roads. Portland has had traffic calming programs in effect since the early 1990s. These engage the community in decisions that result in changes to their streets and neighborhoods.

In traditional traffic engineering, the goal is to get as many cars through the streets as quickly as possible by making streets wide and unobstructed. Traffic calming does the exact opposite: it narrows streets and puts barriers in the way of the driver, forcing him or her to slow down. Traditional methods, such as stop signs and speed limit signs, require enforcement because drivers ignore signs and the laws behind them. In contrast, traffic calming measures are self-enforcing. They rely on physical laws and human behavior to slow traffic.

One drawback to traffic calming is its potential for reducing the speed of emergency vehicles; this creates an additional design challenge for engineers constructing roads for a variety of users.

Speed bumps, humps, and tables

IS IT SAFE? 61

SPEED BUMPS, HUMPS, TABLES

These force vehicles to slow down by raising the pavement three to four inches. Each type creates its own effect on drivers: bumps are steep, while speed humps and speed tables are larger and more gradual. Portland implemented the Residential Speed Bump Purchase Program—unofficially called the "buy-a-bump program"—in response to community pressure for traffic calming measures.

MEDIAN ISLANDS

Median islands give pedestrians a safe place to stand while crossing a wide street. They also shorten the crossing times for pedestrians who might be slower than average. Medians often make drivers more aware of pedestrians and cause drivers to reduce their speed.

Landscaped median island are a main feature of Portland's Woodstock Boulevard traffic calming program. They provide a refuge and a sense of ownership to pedestrians crossing this busy street. Landscaped medians have the added benefit of improving street appearance.

NECKDOWNS, CHOKERS, BULBS

These are the names of curb extensions that narrow the street at mid-block or at intersections. When roads are narrowed, drivers naturally reduce their speed. Pedestrian crossing distances are shorter.

Portlands's Woodstock Boulevard includes curb extensions that double as bus stops. When the bus stops, traffic behind it must halt. No laws, traffic lights, or signage is required.

TRAFFIC CIRCLES

Traffic circles, or smaller versions called round-abouts, slow traffic naturally by compelling drivers to be watchful as they follow the curve.

Safety in Law

Car-centered design is not inevitable. Public groups that advocate infrastructure that allows and encourages walking have helped communities become conscious of the need to include pedestrians in transportation planning. Two important pieces of legislation have significantly changed the pedestrian landscape of the United States.

ADA

The Americans with Disabilities Act of 1991 mandated that people with disabilities have full access to all public facilities in the United States. With design changes such as the inclusion of curb cuts at sidewalk corners, engineers' intensified focus on walkability has provided a benefit to all pedestrians.

ISTEA

The 1991 Federal Transportation bill known as the Intermodal Surface Transportation Efficiency Act, or ISTEA, included the first effort to integrate pedestrian and bicycle traffic into existing transportation modes. The act, which included funding for a pedestrian and bicycle planner in every state, was a significant step towards equalization of the pedestrian-automobile balance. It required states and metropolitan areas to develop long-range transportation plans that include pedestrians and bicyclists.

Portland has adopted a comprehensive transportation plan. It includes a Pedestrian Master Plan that takes into account a voluntary Citizen Advisory Committee appointed by the Commissioner of Traffic. The committee provides advice on the best use of limited resources for pedestrian projects.

> "What's most important is planning for pedestrians: planning for sidewalks, shoulders for suburbs, and new developments that take pedestrians into account."
>
> —Transportation engineer, CHARLES DENNY

Walkability

Streets can change only if governments, the public, and civil engineers change their thinking. As in all infrastructure issues, any of these groups could take the lead.

Increased walkability means roads and communities that don't seem to force people into cars. Walkability means a safer environment for pedestrians.

Walking is a still viable form of transportation. As in all infrastructure issues, public education is essential.

More engineers are thinking about pedestrian-oriented design because of pressure from communities and because professional organizations, including the American Society of Civil Engineers, the Transportation Research Bureau, and the Institute of Transportation Engineers, are addressing these issues.

Traffic engineers refer to "safety-conscious planning," which means designing roads and neighborhoods with safety as a priority. They are recognizing that sometimes the balance must tip in favor of pedestrians and bicyclists.

The tallest! The deepest! The widest! The heaviest!

Outsized physical dimensions have long dominated presentations of civil engineering. Infrastructure is big, but it is also invisible. Infrastructure is public, but but it is also personal.

Why So Big?

Infrastructure serves the public as it reflects public values. How? Each individual shapes infrastructure, through participation in or disengagement from the institutions that determine the policies of the nation. At the same time, each person's lifestyle determines the future of infrastructure.

Although infrastructure is huge, to many people it is also invisible—perhaps because it is everywhere. When a new civil engineering work appears in the everyday landscape, however, the public no longer takes it, or the essential service it provides, for granted. At that point, the size celebration may stop and an awareness of the relationship between civil engineers, infrastructure, the community, and individuals, begins.

Who Is Us?: Community   70

In Tower: Structure   75

iMagazine: Infrastructure   78

NimbyWeek: Backyard   80

WHO IS US?

MY COMMUNITY

INSIDE

CIVIL ENGINEERS LEAD THE PLANNING, DESIGN, CONSTRUCTION, AND MANAGEMENT OF THE INFRASTRUCTURE THAT ALLOWS A COMMUNITY TO EXIST.

INFRASTRUCTURE DEFINES COMMUNITY.

```
#BXBDBCL **************3-DIGIT 127
#70SYT 1133/ /VAN99# 1   OCT02 11 PMM
KAREN SMITH                  P65-179
72 VAN TUYL RD                    AF
POND EDDY PA  22770-5808
```

- WHAT IS COMMUNITY?
- INDIVIDUAL RIGHTS OR COMMON GOOD?
- WHOSE INFRASTRUCTURE?

"Community" may refer to all the people who live within a particular area. It may refer to people who are part of a network of social relationships—individuals who work or play together. "Community" may refer to people who share resources, such as a reservoir or a power station. Which communities does a particular part of the infrastructure serve? Which communities pay? Who benefits?

Communities overlap other communities. Communities encompass others like a set of Russian nesting dolls. The actions of the members of a single community can therefore affect a region, a state, a nation, and the world.

"

There is no community. We are only a collection of individuals.

They never used to care about what we thought.

Every neighborhood
contains multiple communities,
each with its own interest.

The population keeps growing, but our resources are limited.

Do we need wider roads or improved mass transit?

How are we going to afford what the community needs?

Should we permit the construction of a new subdivision?

Paved roads are not enough. We must be able to drive at will through natural landscapes, forests, and parks.

Protection of the environment is the foundation of community.

There's nowhere to walk to.

Americans build big.

We never know how decisions really get made.

Why don't all communities have excellent educational facilities?

WHY SO BIG? 73

Were Americans to write a Bill of Rights for the 21st century, the right to bear cell phones would appear first. This new development in freedom of speech depends on a vast network of structures and cables. From cellular towers to satellites, civil engineers contribute to the expertise that makes it possible.

Height

During the 20th century, the rise of radio required broadcast towers. During the 1950s and 1960s, AT&T built a national network of microwave towers for long-distance transmission; most have been replaced by cable that has the capacity required by the Internet. Today, the 50 states are subdivided into thousands of small zones called *cells*. Each requires a tower to link cell-phone carrying Americans to the communication network.

Design

There are three types of steel towers: the guyed tower, held in place by cables; the lattice tower, a framework of bars in triangular components; and the monopole, a tube. These lightweight, rigid structural types have long been designed by engineers. Today, however, much tower fabrication has been automated. Standard elements are combined to fit each case.

The factors that determine tower choice are factors that determine all infrastructure design. Designers take into account environmental conditions including soil type and earthquake risk, wind loads, and surrounding land use. Equally important are requirements concerning speed of construction and the amount of money available for the project.

iMagazine

Vol. 150, No. 6 November 2002   Infrastructure Magazine International

The Infrastructure of Water

The water collection, treatment, and supply infrastructure designed by civil Engineers allows Americans to take their baths and lawns for granted. During the 19th century, the public celebrated the new infrastructure of water because it allowed the United States to survive and grow.

Waterworks is a term that describes civil engineering facilities that collect, treat, store, and distribute water.

Cities such as Detroit and Chicago once built lavish waterworks. Many communities constructed facilities that celebrated the infrastructure of water and served as tourist attractions. Refined architecture and elegant landscape design acknowledged civil engineering as a basis of civilization and a source of enormous pleasure.

"Water supply was the first important public utility in the United States and I can scarcely conceive a service that demonstrated its indispensability to growth..."

—Historian MARTIN V. MELOSI, *Sanitary City*, 2000

"Infrastructure is both a consequence and a cause of urban development."

—DOUGLAS C. HENTON and STEVEN A. WALDHORN, "Future of Urban Public Works," 1984

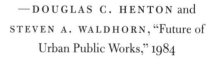

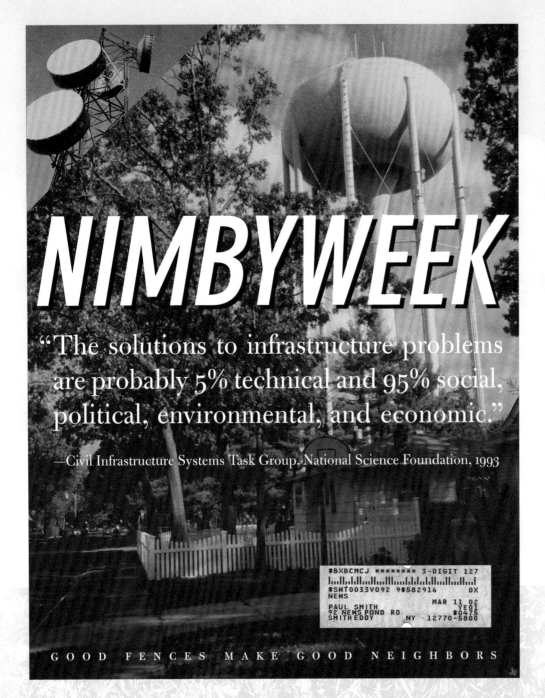

Not in My Backyard

The NIMBY—not in my backyard—phenomenon grew as a result of a series of environmental and public health disasters. These events shook public trust in government and corporate leaders, engineers and scientists.

Although today's civil engineering infrastructure provides a quality of life and a level of safety that would astonish past generations, some Americans are wary of new engineering works. The "wireless revolution," for example, rests on a global infrastructure that includes cables, satellites, and the towers that are conspicuous reminders that most Americans want cell phones and high-definition television. Yet hundreds of towns have banned the construction of towers.

"While almost everyone wants instant, unfettered communications over fat pipes, many people want the associated transmission towers to be located only on the moon, or at least in someone else's backyard where they're out of sight but not out of range."

—*InternetWeek* editor WAYNE RASH

"No engineering problem is without its cultural, social, legal, economic, environmental, aesthetic, or ethical component, and any attempt . . . to approach an engineering problem as a strictly technical one will be fraught with frustration."

—Civil engineer and historian HENRY PETROSKI

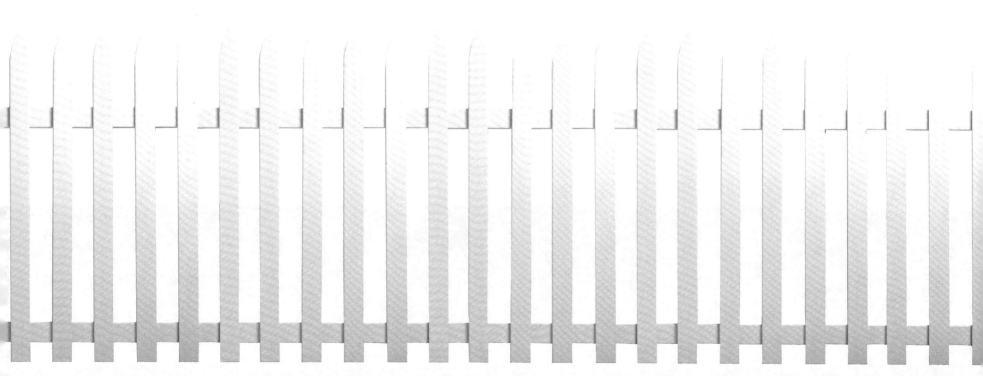

Nimby Logic

What determines the nature of Americans' engagement with their infrastructure—and each other?

CONTROL. Some Americans may feel a loss of control in their lives. They are willing, however, to take enormous risks with their life and health, as long as they feel they are in control.

DISAPPOINTMENT. Public pronouncements about the promise of technology have sometimes been exaggerated.

ENTITLEMENT. Some Americans feel that contributing to the public good is not a requirement of citizenship.

ISOLATION. Americans have left urban areas and mass transit, increasing their isolation and loosening the bonds of community.

SENSATIONALISM. The mass media frequently exaggerate technological mishaps or do not present them in a balanced context.

EDUCATION. The level of technology education in schools is inadequate. Americans of all backgrounds and educational levels lack basic knowledge about the planning, design, construction, and maintenance of the world around them.

POLITICS. Some public leaders plan only for their term of office, not for the long term.

COMMUNICATION. In the past, public officials and civil engineers often did not adequately communicate with the public.

Cellular towers in Georgia, Illinois, New York, and Virginia

DISCRIMINATION.
Until recently, poor Americans and minority groups were provided with an inferior quality of infrastructure.

DISASTER.
Industrial disasters and inconsistent government oversight have made Americans cautious.

MYTH.
Misconceptions about "freedom" and "free enterprise" have led some Americans to believe that their lives are not shaped by the government that has always regulated the activities of businesses and individuals.

LEADERSHIP.
Americans are very concerned about environmental protection. Actions and statements of public leaders have made Americans fearful for future generations.

PASSIVITY.
Some Americans prefer to blame someone else or look for conspiracy rather than attempt to understand and constructively participate in society.

RESEARCH.
When trying to understand complex issues, some people rely on the web and television as their primary research sources, instead of using them as supplements.

CONFUSION.
Public leaders often do not make clear the interrelationships between public and private. They may not present the relationship between who benefits and who pays in the short and long term.

PRIVILEGE.
Infrastructure has allowed Americans to enjoy a quality of life so high that some believe that it comes without sacrifice or sharing.

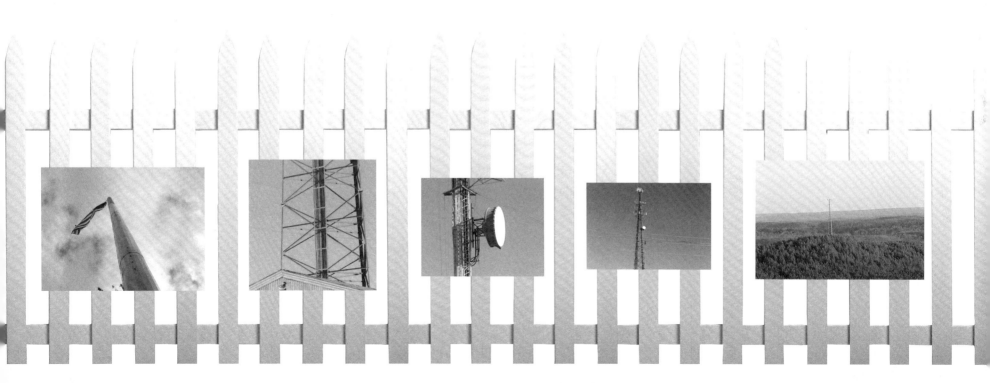

Jackson Heights, New York

IS IT AVAILABLE?

yes WE'RE OPEN

Water

Is It Available?

Convenience	88
Dependability	91
Quality	99
Safety	102
Sustainability	104
Cost	106
Civil Engineers	107
Security	108
Community	109
Disposal and Treatment	111

Without the water running through the tap, offices and factories would close. Home would not be quite as sweet. Still, your body might survive... for a week.

Dependable, safe, cheap, and secure water is the basis of our communities, our businesses, and our pleasures. If this sounds like an exaggeration, that's because we often take our most precious resource for granted—daily, in a cup, with coffee, milk, and sugar.

Beginning in the 1840s, New York City began riding a wave that washed glasses in dockyard pubs and high-rise watering holes. Today, residents of the five boroughs continue sipping and flushing like there's no tomorrow. How do they do it? The infrastructure of availability.

86 ME, MYSELF AND INFRASTRUCTURE

Coffee

Americans consume 45 million cups of coffee per day. Most of each serving is water. Water, moreover, keeps the cups clean. Large quantities of water are required for the irrigation of coffee trees and the processing of beans.

Hamburger

Water consumed in the production of a quarter-pound hamburger:

BEEF (4 OZ.):	616 gallons
CHEESE (1 OZ.):	56 gallons
BUN (2 OZ.):	25 gallons
KETCHUP (1 OZ.):	3.2 gallons
TOMATO (1 OZ.):	1.8 gallons
LETTUCE (1 OZ.):	1.3 gallons

SOURCE: *Stuff: The Secret Lives of Everyday Things* (1997)

Bagel

An amalgam of wheat, yeast, shortening, sugar, and malt, the bagel is a New York staple. About 108 gallons of water are required to produce the 2/5 of a pound of flour that is kneaded, along with an additional cup of water, into a circular shape. And then the baker boils the dough in water before placing it in the oven.

CONVENIENCE

Order. Pay. Sip. *Oops!* Wipe.

The convenience provided by the infrastructure allows immediate cleansing.

Once upon a time, water meant working, waiting, and worrying. If you needed water,
you had to fetch it or have it carried to you.

Before Infrastructure

During the 1700s, New Yorkers knew that they lacked an adequate water supply. Politicians spent decades debating whether a new water system should be public or private. In the interim, New Yorkers relied on common wells, water deliveries from Brooklyn, and the 48-acre Collect Pond, which eventually became known as "the common sewer."

Pumping Tea Water

The Tea Water Pump once supplied the best water in Manhattan. Its source was a spring beneath a private well; its product soon became the standard for high quality. A network of labor and commerce arose around the pump. By 1774, Tea Water Men brought water to 3000 New York homes. The pump supplied most of the city.

Water is Heavy

Before the Croton Aqueduct, New Yorkers got water from wells and rooftop barrels. This usually required hauling heavy buckets of water long distances. One gallon of water weighs more than eight pounds.

IS IT AVAILABLE? 89

No Infrastructure... No Water

Since the enactment of the North American Free Trade Agreement, more than five million Mexicans moved to the border for employment in corporation-owned factories called *maquiladoras*. At least 40% of this exploding population lives in communities that lack infrastructure including water supply and treatment plants and sewer systems.

The divining rod, still used in many parts of the world, is a forked stick held in both hands while wandering in search of water. A force pulls the stick toward the ground at the spot under which water lies. This method, considered a fraud, is an option for populations lacking an infrastructure that makes water available.

Water supply, Nogales, Mexico, 1998

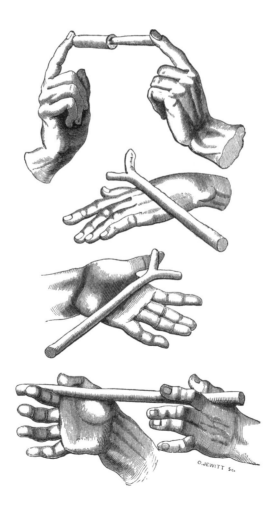

DEPENDABILITY

The New York City water system is one of the greatest achievements in urban history.

The dependability and efficiency of this network of reservoirs, tunnels, and pipes made it a model throughout the United States.

Civil engineer John Jervis designed the system to flow from the Catskills to local distribution points with only the help of gravity—the same force that keeps coffee in its cup. A gravity system does not rely on mechanical pumps, which require substantial resources to operate and maintain.

Left: John Bloomfield Jervis (1795–1885)

Early Plans

War, economics, and politics thwarted plans for a water system in New York City. In 1798, the New York Assembly granted Aaron Burr $2 million dollars for his new Manhattan Company. It was formed in order to create a citywide water system, but it became a platform for Burr's financial goals and political ambitions. For thirty years, the Manhattan Company refused to cede its rights or act upon its promise to build a water system.

Croton Aqueduct

When the Croton Aqueduct opened in 1842, it provided 12 million gallons of fresh water a day to a city with a scarce, contaminated water supply.

The Croton system encompassed a 100-foot stone-masonry dam, a 400-acre lake, 41 miles of aqueduct, 16 underground tunnels, the Sing Sing Kill and Harlem River bridges, a receiving reservoir (now Central Park's Great Lawn), and a distributing reservoir at 42nd Street (current site of the New York Public Library). Engineers decommissioned the 42nd Street and Central Park Reservoirs in 1890 and 1925, respectively. The Old Croton Aqueduct continues to bring water to Ossining.

Above: construction of the first Croton Aqueduct, 1890s

The Murray Hill Distributing Reservoir at 42nd St., New York, 1893

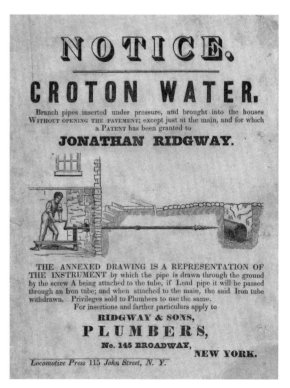

Advertisement for Plumbers, no date

New Croton Aqueduct

In 1869, New York City's 942,000 citizens consumed 77 million gallons of water each day. In the 1880s, an exploding population and a period of drought led civil engineers to recognize that the Croton Aqueduct would not be adequate if the city was to survive and grow.

Civil engineers designed the New Croton Aqueduct to supply New York with three times the amount of water provided by the original aqueduct.

The reservoir created by the new dam engulfed the old dam. Water in the new 35-mile-long aqueduct flowed to the Jerome Park Reservoir in the Bronx, where it was tunneled 300 feet below the Harlem River for its trip south.

IS IT AVAILABLE? 93

High Bridge

John Jervis and James Renwick, Jr., designed the High Bridge, through which water flowed in two pipes from the Bronx to Manhattan over the Harlem River. High Bridge, 1,450 feet long and 135 feet high, was notable for its 15 stone arches. In 1923, engineers replaced a portion of the bridge with a steel arch that was less of a barrier to navigation; several of the original arches survive on the Bronx side. Completed in 1848, High Bridge is said to be the oldest bridge in New York City.

Catskill and Delaware

Within one year of the opening of the New Croton Aqueduct, water consumption climbed from 110 to 165 million gallons per day. The 1898 incorporation of Queens, Brooklyn, and Staten Island into New York City placed an additional burden on the water supply.

This time the city turned to the Catskill watershed, which encompasses 571 square miles 100 miles north of New York City. Built between 1907 and 1917, the Catskill System is composed of the Ashokan and Schoharie Reservoirs. The Catskill System today supplies 40% of New York City's water.

The Delaware reservoir system, built 1937-65, was the next addition to New York's water infrastructure. Its four reservoirs (Roundout, Neversink, Pepacton, and Cannonsville) provide approximately 50% of the city's water.

Construction of the Catskill Aqueduct, 1913-1915

Background: High Bridge

Tunnels Nos. 1 and 2

City Tunnel No. 1, constructed between 1911 and 1915, brings water from the Hillview Reservoir in Yonkers to distribution systems in the Bronx, Manhattan, Brooklyn, Queens, and Staten Island. It is 18 miles long.

City Tunnel No. 2, completed in 1933, also begins at the Hillview Reservoir, and brings water to distribution systems in the Bronx, Queens, and Brooklyn. Both tunnels have a capacity of one billion gallons a day.

City Tunnel No. 3

Construction workers are currently blasting a third tunnel through bedrock as deep as 800 feet below the streets of New York. City Tunnel No. 3, which measures up to 24 feet in diameter, will deliver a maximum of 1.2 billion gallons of water daily. It will allow civil engineers to inspect and make repairs to Tunnels No. 1 and No. 2 for the first time. In addition, the tunnel will maintain water availability in the case of an emergency. Although Water Tunnel No. 3 will not be completed until 2020, a portion was put into service in 1998.

There is an urgent need for the maintenance work that City Tunnel No. 3 will allow. The Delaware Aqueduct tunnel, which supplies the New York City region with 50% of its drinking water and is one of the largest in the world, may be leaking as much as one billion gallons of water a month.

Construction of City Tunnel No. 3, Queens, New York, 1999.

Gravity

Much of the dependability of New York's water supply is the result of the role of gravity in its design.

Hillview Reservoir collects water at 295 feet above sea level. From there the water drops into giant pressure tunnels that run as deep as 800 feet below ground. Seeking to attain the same level as the reservoir, the water piped into the city can reach a sixth-floor apartment. Because this system does not require pumps, it is cost-effective and environmentally friendly.

Pressure

New York City's water is distributed by a network of 6,181 miles of water mains, regulated by more than 88,600 valves. Valves control the amount of water that flows through pipes. Because the system is gravity driven, water pressure must be decreased, rather than increased.

Pumping Stations

Under normal conditions water in a gravity-fed system can flow upwards unaided for 30 feet—the furthest that air pressure can lift water. Engineers can increase this number by raising the level of the water source or by employing pumps.

Pumps have always been a key to water supply. Historically, animals or humans provided the power. Steam, and eventually, electricity replaced them. Many cities need to rely almost entirely on pumping stations; Los Angeles, for example, uses 70 pumping stations.

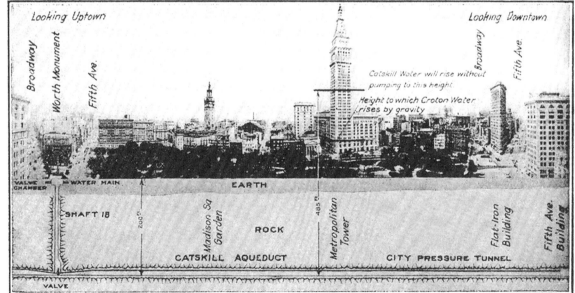

Water pumping station, Buffalo, New York

IS IT AVAILABLE? 97

Water Towers

Water towers provide storage and regulate water pressure for homes and businesses. During peak hours, such as weekday mornings, water towers keep pressure constant even if everyone flushes at the same time.

In cities, rooftop tanks provide water for taller buildings. Pumps lift water into the tanks; gravity causes the water to descend.

Water tanks are hidden at the top of skyscrapers. Valves reduce pressure so that pipes do not rupture. Fire sprinklers require a separate supply, so their tanks are located within the building.

QUALITY

Like the cities and suburbs it flows into, tap water is an engineered product.

By designing a system of storage, treatment, distribution, and monitoring, civil engineers "invented" a high-tech fluid that appears natural.

Drinking water was not even a beverage concept in early New York. Quality was so poor that the thirsty preferred alcoholic drinks. If they added water to the punch, it was in the form of boiled tea.

Coffee shop culture began when five coffee houses opened circa 1760, bringing the total number to seven.

Water of Yore

Once upon a time, it was safe to drink directly from lakes and streams. This is how it worked. Rain and melted snow percolated through layers of earth until, at groundwater level, it was free of pollutants, bacteria, and sediment. It then flowed into the voids between rocks, where New Yorkers could reach it via wells. Or it flowed into lakes and rivers, and could be directed to reservoirs that stored the clean liquid for a thirsty populace.

" There's nothing new in town, except the Croton Water, which is all full of tadpoles and animalculae, and which moreover flows through an aqueduct which I hear was used as a necessary by all the Hibernian vagabonds who worked upon it…Post drank of some of it and is in dreadful apprehension of breeding bullfrogs inwardly."

—Literary critic GEORGE TEMPLETON STRONG, 1842

Pollution

In 1832, the 330,000 inhabitants of New York City collected their water from wells, barrels on rooftops, or had it delivered by cart. Since they disposed of waste in open canals and streams, the city's underground aquifers and wells became very polluted. Overflowing cesspools and open sewers were the perfect breeding ground for infectious diseases such as yellow fever, diphtheria, mumps, typhoid fever, and cholera. In 1832, cholera killed an estimated 3,500 New Yorkers.

Water Treatment

The Croton Aqueduct provided New Yorkers with an abundant supply of water, but cleanliness was still a concern. At the beginning of the 19th century, engineers introduced filtration, treatment, and improved sewerage practices.

Cost, ease, and efficiency determine the kind of water treatment system a civil engineer selects. Before it reaches the tap, water is shaken, stirred, flushed, and filtered, generally through sand. Often, the water is treated with chemicals, which remove sediment and harmful bacteria.

Code Letters	Nature of Odor	Description or Probable Cause of Odor
A	Aromatic (Spicy)	Such as odor of camphor, cloves, lavender, and lemon
Ac	Cucumber	Such as odor of the protozoa known as Synura
B	Balsamic (Flowery)	Such as odors of geranium, violets, and vanilla
Bg	Geranium	Such as odor of the algae known as Asterionella
Bn	Nasturtium	Such as odor of the algae know as Aphanizomenon
Bs	Sweetish	Such as odor of the algae known as Coelosphaerium
Bv	Violets	Such as odor of the protozoa known as Mallomonas
C	Chemical	Such as odors due to industrial wastes or chemical treatment
Cc	Chlorinous	Such as odor of free chlorine
Ch	Hydrocarbon	Such as odors of oil refinery wastes
Cm	Medicinal	Such as odors of phenol or iodoform
Cs	Sulphuretted	Such as odor of hydrogen sulphide
D	Disagreeable	Pronounced unpleasant odors
Df	Fishy	Such as odors of the protozoa known as Uroglenopsis and Dinobryon
Dp	Pigpen	Such as odor of the algae known as Anabaena
Ds	Septic	Such as odor of stale sewage
E	Earthy	Such as odor of damp earth
Ep	Peaty	Such as odor of peat
G	Grassy	Such as odor of crushed grass
M	Musty	Such as odor of decomposing straw
Mm	Moldy	Such as odor of damp cellar
V	Vegetable	Such as odor of root vegetables

Above: Odor list for water testing, W.A. Hardenbergh, *Water Supply and Purification* (1952); *Below:* Water filtration plant.

Public Health

New York's source is the remarkably clean runoff from upstate mountains. Although little filtration is needed, water is treated with small amounts of fluoride for dental health. Chlorine, a disinfectant, is added to kill bacteria that cause illnesses including typhoid, cholera, and hepatitis.

Chlorine treatment was "probably the most significant public health advance of the millennium," according to *Life* magazine.

Human Existence and Human Health

Water consists of one oxygen molecule and two hydrogen molecules. The average human being can survive as much as a month without food, but no more than a week without water. Although 75% of the earth's surface is covered with water, only 1% is available for drinking. The human brain is 70% water.

Taste Test

Taste, color, odor and turbidity were the earliest standards for measuring water quality. At the beginning of the 19th century, advances in the understanding of germs, along with the introduction of the microscope, helped scientists to establish three categories of bacteria: "natural," "soil," and "intestinal and sewage" bacteria. They believed that the latter caused diseases such as cholera and typhoid fever.

Quality Test

Using the more than 1,000 sampling stations throughout the city, the New York Department of Environmental Protection tests almost 40,000 water samples each year. The DEP performed 752,000 analyses on 39,100 water samples in the year 2000. It monitors several indicators of water quality, including pH and contaminant levels.

In 1992, the city instituted a pathogen monitory program at the Catskill and Delaware Reservoirs, to test for Giardia and Cryptosporidium, microorganisms that cause intestinal illness.

SAFETY

Although Manhattan is surrounded by water, devastating fires repeatedly put the island at risk.

Just as engineers designed convenience, dependability, and quality, they helped create the infrastructure of safety during the last two centuries.

Great Fire of 1835, as seen from Long Island

The Great Fire of 1835

The Great Fire of 1835, which destroyed 600 buildings, was the most prominent of the catastrophes that led to the Croton system. New York's rudimentary fire department was unable to stop the flames that engulfed the city. Their water froze in water carts before it could be pumped. Damages were estimated at $18 to $20 million dollars, and New York's insurance industry went bankrupt.

Destruction

Before dependable, citywide water systems, fires were a constant threat. Chicago lost more than 18,000 buildings in 1871. Three decades later, a fire destroyed more than 1500 buildings in Baltimore. These cities were dependent on heroic firemen wielding buckets of well water.

Firemen

Organized firefighting began in New York City (then New Amsterdam) in 1648 with a fire watch composed of eight wardens. More firefighters soon patrolled the streets with buckets, hooks, and ladders. The first bucket brigade was organized in 1658, with members using leather buckets filled with water. New York imported the first hand-pump fire engines from England in 1731. In 1859, New York's fire department began to use steam powered fire engines.

Fire Hydrants

Fireman George Smith is credited with inventing the fire hydrant in 1817. New York installed its first hydrants in 1830, but Smith's invention was not integrated into the water system until almost a century later. The importance of fire hydrants was highlighted by the terrorist attack on the World Trade Center in 2001. Hydrants and pipes destroyed in the collapse, both above and below ground, created an obstacle for firefighters. Old-fashioned availability of water was essential to controlling the fire.

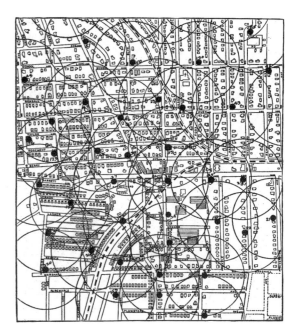

SUSTAINABILITY

It's one of the great paradoxes of infrastructure: the more you have the more you need.

In New York, the increasing supply of water increased demand. The population grew explosively after the completion of the Croton Aqueduct, from 20,000 in the 1840s to 515,547 in 1850. Moreover, the quantity of water used by each person increased. Today, the water system supports 19 million.

Home

Americans use 24 billion gallons of water at home each day. They take more showers than any people on earth. How long can this lavish lifestyle last?

DAILY DOMESTIC WATER USE	GALLONS	PERCENT
Total per person per day	168.00	100%
FLUSHING THE TOILET	48.72	29%
SHOWERING	35.28	21%
WASHING MACHINE	35.28	21%
USING THE TAP	20.16	12%
BATHING	15.12	9%
LEAKY TOILETS	8.40	5%
DISHWASHING	5.04	3%

SOURCE: U.S. Environmental Protection Agency

Industry

Of the 402 billion gallons of water consumed each day in the United States, 57% is used to irrigate crops, support livestock, produce electricity, and support industry and commerce.

Irrigation-livestock: 40.9%
Thermoelectric 38.7%
Domestic-commercial: 12.2%
Industrial-mining: 8.2%

The End of Water?

Water is a finite resource.

States including New York, Massachusetts, Illinois, Kentucky, California, and Oregon, are experiencing water shortages. This situation promises to worsen. Some experts blame global warming for reduced surface water in lakes and rivers. At the same time, consumer and agricultural demands are depleting the underground rivers and lakes that supply 60% of the nation's water.

Civil engineers are responding to water scarcity in a variety of ways. Some are seeking to build new pipelines to distant sources. Some are developing new technologies for monitoring water loss. Others are exploring water treatment methods that will render saltwater and wastewater potable. Still others are engaged in making treated wastewater available for gardens and lawns.

Conserve

Through the use of meters and leak detection, the New York City area has significantly decreased its water consumption. Today, residents consume 1.3 billion gallons of water per day, down from 1.8 gallons in 1989. One strategy was the introduction of hydrant spray caps, which reduce water flow to 25 gallons a minute. Otherwise, hydrants can lose millions of gallons each day.

Flush

Spurred by a rebate program offered by the Department of Environmental Protection, New Yorkers began in 1994 to replace their water-guzzling toilets with low-flow toilets. The city offers rebates to residents who install the new toilets. Low flush toilets have the benefit of using 1.5 gallons of water per flush instead of the traditional 3.5 gallons. Between 1994 and 1997, 1.33 million toilets were replaced, resulting in a savings of up to 90 million gallons per day.

Play

26.4 million Americans enjoy golf on 16,365 courses nationwide. They play on lawns whose maintenance puts stress on limited water sources. Recently, golf associations and environmental groups, working together with civil engineers, established a set of environmental principles for golf courses.

One solution is to engineer courses so that they help to replenish underground reservoirs. In Sun City, Arizona, storm water is diverted into trenches located underneath a golf course. Then it is purified naturally as it seeps through the earth prior to entering the aquifer.

COST

The water supply infrastructure allows us to see water as bountiful and free. It *is* amazingly cheap.

BOTTLED WATER: 1,000 gallons........ $4,000   TAP WATER: 1,000 gallons....... $1
WATER FOR INDUSTRY AND AGRICULTURE: 1,000 gallons........ 10¢

Comparison Shopping

CROTON SYSTEM (1842): $13 million

HIGH BRIDGE (1848): $963,427

CITY TUNNEL No. 3 (2020): $6 billion

SOURCE: Supermarket in Queens, New York, 2002

Guest Check

CHECK NO. 2651

PRICE PER PINT	
Snapple	$1.00
Coca-Cola	$0.80
Milk	$1.00
Tropicana Orange Juice	$0.83
Motts Apple Juice	$0.70
Gatorade	$0.85
Ocean Spray Cranberry Juice	$1.10
Tap Water	$0.0001

"Just Say No To H20"

Some corporations have profited from the bottled-water craze by selling treated tap water. In an effort to dissuade the public from drinking tap water, a leading soft drink manufacturer recently developed a program for wait staff. Its "Just Say No" slogan encouraged the steering of customers from tap water to bottled beverages.

CIVIL ENGINEERS

Public Advocates

Civil engineers have always functioned as advocates for technologies they believe improve the quality of life. They were key advocates for public water systems.

A public that does not want to face difficult choices or seemingly unlikely risks sometimes believes the analyses of civil engineers are exaggerated. At other times, the public perceives engineers as conservative. The professional approach of civil engineers is shaped by the knowledge that their work means life or death for each individual.

ASCE 150th

In 1852, twelve engineers met to establish what became the American Society of Civil Engineers. They assembled in the New York City office of the Croton Aqueduct—headquarters of a technological achievement that formed the foundation of urban America.

John Bloomfield Jervis (1795-1885)

From 1836 to 1846, John Jervis was in charge of the construction of the Croton Aqueduct. A participant in the building of the Erie Canal, Jervis worked as designer and manager of reservoirs and locomotives. When ASCE was founded in 1852, he was directing railroad construction in the Midwest.

Alfred W. Craven (1810-1879)

The first ASCE meeting took place in the office of Alfred W. Craven, then chief engineer of the Croton Aqueduct. Craven supervised the construction of a reservoir in Central Park, surveyed the Croton watershed, and improved sewer systems in Manhattan and Brooklyn.

James Laurie (1811-1875)

James Laurie's experience in founding the Boston Society of Civil Engineers was useful in New York, where he became the first president of ASCE. Laurie was an experienced railroad engineer whose work included the surveying and design required for the construction of tunnels, bridges, and dams.

SECURITY

Since the survival and prosperity of a nation hinges on the availability of water, infrastructure is a target in times of war.

War

Measures to protect the water supply date back to the Civil War. In 1864, indications of a Confederate plot to destroy several New York buildings prompted Union General Benjamin F. Butler to bolster forces around the reservoir. Thirty-five hundred troops protected the High Bridge and Croton Aqueduct from sabotage.

World War II defense included increased security measures. Cold War concerns led to round-the-clock patrols. The government established mobile laboratories that could analyze water in case of attack.

Nuclear, Biological, and Chemical Threat

The attack on the World Trade Center in 2001 meant that, once again, New Yorkers became aware that a terrorist could contaminate the water supply. There was an increase in security and surveillance, as well as restricted access to land adjacent to reservoirs. New York City commissioned the Army Corps of Engineers to design a $30 million security system in order to better protect their source of water.

COMMUNITY

No city could have grown without its water supply infrastructure. Everyone's lives and communities are water-based.

Available, healthy water depends on community cooperation. But who is the community? Who is *us*?
The answer to this question determines how and if civil engineers can maintain the infrastructure that shapes everyone's lives.

Pride

Once, waterworks and water towers were points of civic pride. Detroit's Waterworks Park, for example, was a popular attraction. Today few Americans are aware of the infrastructure behind each glass of water.

The world's largest catsup bottle, in Collinsville, Illinois, holds 100,000 gallons of water.

This one million gallon peach in Gaffney, South Carolina is 135 feet tall.

The Chicago Water Tower, a city landmark built in 1869, was designed to resemble a medieval castle.

IS IT AVAILABLE? 109

"If there has been one common denominator in the city's multitudinous water projects, it has been the conflict between individual property rights and the needs of the masses."

—DIANE GALUSHA, *Liquid Assets*, 1999

Sacrifice

Every urban water system requires sacrifice in surrounding communities. The construction of the Croton system flooded thousands of acres of farmland. Some landowners took legal action. The State of New York determined that access to water took precedence over individual ownership of land.

The New Croton Reservoir flooded four communities: Katonah, Golden's Bridge, Purdy's Station, and Croton Falls. The Ashokan Reservoir, completed in 1915, flooded eight towns. Several reservoirs, such as Neversink, are named after the towns they covered.

Cooperation

As populations grow, suburbs expand, and agricultural land becomes increasingly industrialized, New York City's once pristine waters are becoming polluted by fertilizers, pesticides, litter, paint solvents, and livestock waste.

In order to avoid building an expensive filtration plant, a memorandum of agreement was put into effect in 1997. It requires residents and businesses on upstate watersheds to abide by a series of strict land-use laws. The city has enlisted the cooperation of dairy farmers whose livestock has been a source of contamination. New York will purchase land abutting reservoirs and streams from willing sellers, and improve roads and bridges to reduce run-off.

"I HAVE TO EXCUSE MYSELF FOR A MOMENT"

New York's water disposal and treatment infrastructure is the product of nearly two centuries of civil engineering.

1.349 billion gallons of sewage flow through New York City each day. Its water treatment system is composed of 14 treatment plants, 89 wastewater pump stations, nine laboratories, eight sludge dewatering facilities, and three inner-harbor vessels that transport sludge between facilities.

Integrated Infrastructure

A water supply infrastructure that includes waste treatment is essential for the health of the public and the environment. Without a system of removal, waste is dumped back into the water supply. This leads to a contamination that poses an acute threat to marine and animal life.

Filth

Scientists in 1830 estimated that New Yorkers dumped 100 tons of human excrement into the soil each day. Privy vaults and cesspools allowed filth and contagion to seep into the city.

Water Closet

Water closets came into vogue in New York during the mid-1800s. As soon as cities had access to running water, the use of these "modern" appliances soared. By 1880, about one-third of New York's households had water closets.

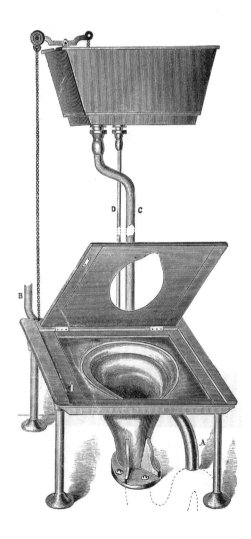

Sewered Cities

At the beginning of the 20th century, sewerage lines were commonplace in New York and had proven crucial to public health and sanitation.

The 1920s witnessed a proliferation of sewered cities, which by 1940 were universal. Cities were still growing, however, and this meant that many sewerage systems needed to be expanded.

Activated Sludge

The Milwaukee Sewerage Commission decided in 1919 to use the biological process of activated sludge, one of the most significant developments in sewage treatment of the last century. The Jones Island Sewage Treatment Facility in Milwaukee, Wisconsin, was a pioneer in the use of the process, developed in 1913 by a scientist in Birmingham, England.

During the activated sludge process, air bubbles pumped into water create a sludge composed of organic bacteria. These bacteria eat the nutrient rich solid waste suspended in the water. After the waste settles, the treated liquid is usually used in agriculture.

Filtering

Before New York's wastewater is treated in its 14 facilities, the Floatable Reduction Program employs large filters to remove floatable contaminants and street litter. Oil and grease are then skimmed off before the water flows to the treatment facilities.

Wastewater treatment begins with several forms of filtration—one of the oldest forms of water purification. While rudimentary forms of filtration have been used for thousands of years, engineers began developing advanced filtration devices in the middle of the 19th century.

Facing page: Plans for a tenement outhouse, Brooklyn, New York, 1906.
Below: Scum collector removes solid material from sewage.

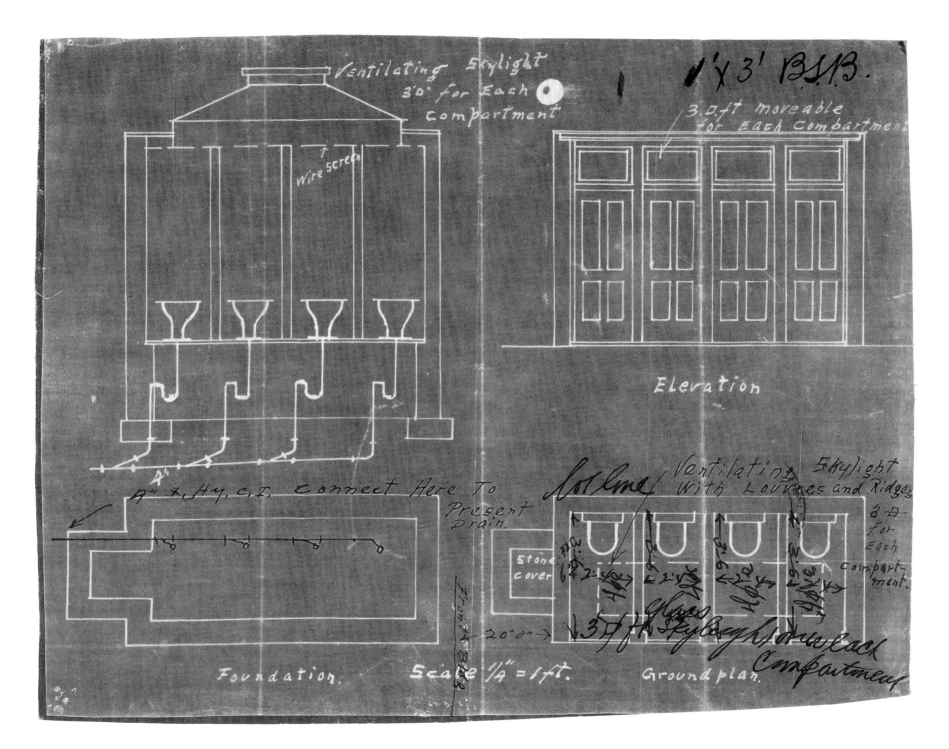

IS IT AVAILABLE? 113

HOW MUCH DOES IT COST?

Microchip fabrication plant

Facing page: Strawberry farm, Santa Clara County, California, 1870s

How Much Does it Cost?

PLACE: Industrial Site 119
MANUFACTURE: Microchip 122
USE: Server Farm 129
DISPOSAL: Cathode Ray Tube 134

Silicon Valley

If there is a house or a tree in the United States, someone decided to build it or let it stand. Houses and trees function within built and natural environments connected by the infrastructure.

The relationship between built and natural determines the real cost of technologies, from highways to espresso makers to the monitors glowing on millions of desks.

A price sticker only hints at the cost of a computer. Manufacturing a microchip, visiting a web site, disposing of unwanted hardware—these activities reveal the actual costs of the information age.

Left: Map of Silicon Valley

Right, top to bottom:
Old Santa Clara County;
Two views of Stanford University;
Prune orchard, Santa Clara Valley, c. 1900

PLACE SILICON VALLEY

Suburb

Silicon Valley is a product of the great migration of the latter half of the 20th century, from city to suburb and from east to west. In search of space, air, and opportunity, Americans abandoned the urban infrastructure built at enormous cost during the previous 150 years.

Placelessness

Silicon Valley is a real place. Real people live and work there. During the Internet boom, promoters used terms such as "placeless," "wireless," "paperless," and "clean" to describe a dreamland inhabited by information industry technologists and consumers. Like all industry, however, cyberspace is earthbound. There are innumerable structures in cities and suburbs, infinite strands of wires, and mountains of paper. There is plenty of dirt. Welcome to Silicon Valley.

Declaration of Independence

"Governments of the Industrial World, you weary giants of flesh and steel, I come from Cyberspace, the new home of Mind."

In 1996, John Perry Barlow, a former lyricist for the Grateful Dead, wrote a manifesto entitled "A Declaration of the Independence of Cyberspace." The Internet, he wrote, "is a world that is both everywhere and nowhere, but it is not where bodies live." He warned the government: "Do not think that you can build it, as though it were a public construction project."

Barlow promoted the immateriality of cyberspace. It consists, however, of physical structures whose environmental costs are only beginning to come to public attention.

"A bit has no color, size, or weight, and it can travel at the speed of light. It is the smallest atomic element in the DNA of information. It is the state of being: on or off, true or false, up or down, in or out, black or white."

—Director of MIT Media Laboratory
NICHOLAS NEGROPONTE, 1995

HOW MUCH DOES IT COST? 119

Fertility

In the early 1920s, the San Jose Chamber of Commerce dubbed Santa Clara Valley the "Valley of Heart's Delight" to advertise its agricultural productivity and healthy living environment. It was said that the fertility of the soil was second only to that of the Valley of the Ganges.

Superfund

The computer industry may seem environmentally clean, efficient, and safe, in contrast to traditional industries. But Santa Clara County, home of Silicon Valley, has the most EPA Superfund sites of any county in the nation. Superfund sites are uncontrolled or abandoned places where hazardous waste has the potential to threaten local ecosystems and populations.

Valley of Heart's Delight

Onion Field, Valley of Heart's Delight

Below: Santa Clara "pioneer" James Kenyon and his residence, c. 1890. Kenyon came to California from Ohio during the 1849 gold rush and bought land west of the town of Santa Clara.

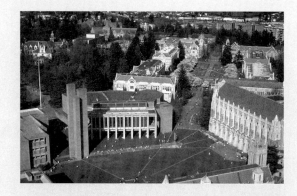

University of Washington in Seattle (*top*) and Carnegie Mellon in Pittsburgh (*above*) are sites of new research centers.

Below: "First building of Silicon Valley," 1953: Varian Associates moved into the newly created Stanford Industrial Park.

University

The history of Silicon Valley is tied to Stanford University, which needed funds to fuel its postwar growth in the 1950s. Stanford offered long-term leases to high-tech companies that would benefit the university. This also kept graduates nearby.

Urbanity

Internet growth has occurred in a relatively small number of countries and urban areas. Ten cities account for 1.5% of the world's population—and 25% of Internet domain names. These are the addresses of individuals, companies, and institutions whose Internet homes operate within civil engineering infrastructure. Just like their real homes.

Creativity

Civil engineers, information technologists, musicians—all innovators work within human networks. Face-to-face contact is usually essential. Meetings—in coffee shops, at conferences, in parking lots, outside nightclubs—are critical.

There is no substitute for being near colleagues and co-workers. That is the reason for the existence of a Silicon Valley.

Proximity

Concentrating individuals in a particular place is a well-known strategy for promoting innovation. Today, Intel is setting up research and development facilities near college campuses in Berkeley, Seattle, and Pittsburgh, in order to take advantage of the faculty and students at these institutions.

MANUFACTURE
MICROCHIP

Civil engineers are responsible for the mining operations, water supply and treatment, transportation, communication and power networks, and factory construction that shape every aspect of information technology fabrication. Microchip, infrastructure, and environment share a close relationship.

Microchip

The silicon microchip is the computer's brain. It is also an economic engine. The chip is small, but its speed and power are enormous. So are the materials and processes behind its creation. The chip, also called an *integrated circuit*, is a semiconductor wafer that includes thousands or millions of resistors, capacitors, and transistors. These components act either as gateways for electrical signals or store energy in the form of an electrostatic field, creating computer memory.

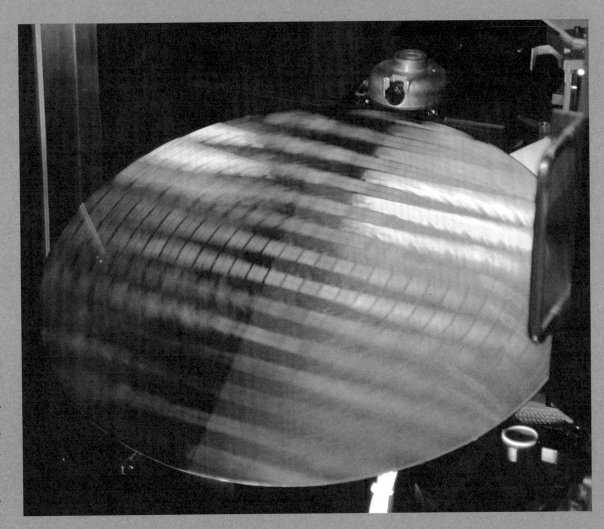

A silicon wafer is cut into hundreds of microchips.

Making Waste

A computer consists of 700 materials and chemicals, many used in the 300-step process of microchip manufacture.

Making the chips for an average computer generates 89 pounds of waste and requires 2,800 gallons of water. Although each chip weighs only three ounces, making chips generates more waste than any other stage of computer manufacture.

Mining

The cost of mining and transporting the raw materials and finished products that end up in a computer or information network—and the construction of the infrastructure required to carry out these activities—is huge.

- Gold connects chips to the computer.
- Copper can increase a chip's power without increasing its size. Virtually all electronic equipment contains copper wiring.

Mining produces millions of pounds of mineral waste contaminated with toxic metals, such as the cyanide used to extract gold from ore.

Mining space is an essential part of cyberspace. Both the benefits and the costs can be found there.

Pollution

In the early 1980s, it was revealed that computer manufacturers had leaked tens of thousands of gallons of toxic contaminants into Silicon Valley's groundwater. The pregnant women of one San Jose neighborhood—recipient of some of the water—had a rate of birth defects and miscarriages 2.5 to 3 times higher than expected.

Gold mine

Clean Room

According to the Intel Museum, "the tiniest speck of dust on a chip could ruin thousands of transistors." To prevent damage to chips from particles of clothing and flakes of skin, workers who make chips must take air showers before stepping into their "bunny suits."

Clean rooms, where chips are manufactured, contain one to five particles per cubic foot of air, thanks to an air filtration system that sucks air through the floor. Hospital operating rooms contain between 10,000 and 100,000 particles. Outside air contains up to one million.

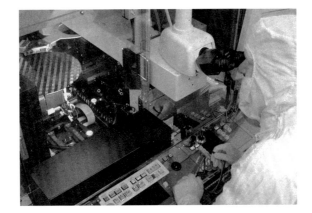

Ultrapure

Microchip manufacture depends on water supply infrastructure. The biggest chip fabrication plants, up to two football fields long, use more than a million gallons of water a day to produce ultrapure water—essential to clean and rinse the silicon wafers that are cut into hundreds of microchips.

Above: Workers in bunny suits
Above right: Working on a wafer
Right: Filtration system in fabrication plant

Water Supply

From the 1850s to the early 1900s, civil engineers put California's water supply infrastructure into place. This system of dams and aqueducts now bridges the gap between supply and demand. Although California's largest water users are located south of San Francisco, most of its freshwater supply originates in the north, in places such as the Sierra Nevada Mountains.

In Santa Clara County, home to Silicon Valley, water demand has exceeded the semiarid area's supply since the early 1940s. Only 50% of Santa Clara County's water comes from local winter storm runoff, captured through a system of dams, percolation ponds, and canals.

Recently, the *San Jose Mercury* rated the construction of the water supply system one of the top ten events that shaped Silicon Valley.

Left: North Fork Dam, Santa Clara County

Below: Sierra Nevada Mountains

Dam

Engineers Michael O'Shaughnessy and Carl Grunsky led the Hetch Hetchy water supply project that included the O'Shaughnessy Dam, completed in 1923. Water collected behind the dam, within Yosemite National Park, travels 160 miles to San Francisco. It is so pure that it requires only basic disinfection. In 1952, Santa Clara Valley started receiving water from this system.

Demand

In 1965, the 44-mile South Bay Aqueduct began bringing Santa Clara County water from the Sacramento-San Joaquin River Delta. In 1987, civil engineers completed the San Luis Reservoir at a cost of more than half a billion dollars. These projects linked Silicon Valley to the larger California State Water Project, whose $1.75 billion bond issue was approved by voters in 1960.

Today, conservation and water reclamation, rather than the expansion of the water supply infrastructure, meet water demand. Continued population and economic growth could cause a shortage within 20 years. Civil engineers will need to analyze the benefits and costs of increasing water supply and present them to an increasingly thirsty public.

> "We are here to mark our progress in constructing the boldest water project in the history of this or any other nation."
> —Governor EDMUND G. "PAT" BROWN, upon completion of South Bay Aqueduct, 1965

Left to right: Hetch Hetchy Reservoir; intake towers, prior to filling of the San Luis Reservoir, 1967; San Luis Reservoir during a drought

Departure

Labor and land costs, as well as new requirements that take environmental impact into account, are driving many electronics manufacturers to locate fabrication plants outside of Silicon Valley. They often build factories where water is scarce and environmental regulations are not as protective. Some new sites are in the desert environments of New Mexico, Texas, Arizona, and Israel.

Conservation

The semiconductor industry has designed chip fabrication plants that monitor water use. These plants, which recycle more than 50% of the water used in chip manufacture, are relatively rare. The new technology is not being installed in older factories because the investment is considered too great.

Sustainability

Given the vibrant role computers play in society, and their environmental costs, is there a way of reaching a sustainable solution to computer manufacture? The industry has funded few inquiries into the environmental and health effects of its activities. It claims that its method of manufacture is the only way. Critics consider their attempts at environmental sustainability modest.

Dedication ceremony, South Bay Aqueduct, July 1, 1965

Innovation

Researchers, such as those at the NSF/SRC Engineering Research Center for Environmentally Benign Semiconductor Manufacturing, which comprises seven universities led by the University of Arizona at Tucson, are developing low-impact methods of manufacturing chips. They have invented ways to imprint circuit patterns on chips, minimizing waste and the use of harmful materials. In addition, they have greatly reduced the need for water in chip manufacture while devising ways to reuse it.

Right: Chip production at the NSF/SRC ERC for Environmentally Benign Semiconductor Manufacturing

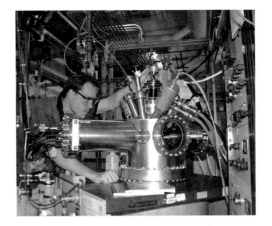

Moore's Law

In 1965, Gordon Moore, co-founder of Intel, predicted that the number of transistors per square inch on integrated circuits would double every year. When Moore made his prediction, the world's most complex chip had 64 transistors. Today, Intel's Pentium III contains 28 million.

Though chips continue to shrink, some researchers believe that size and speed limits of silicon chips will be reached. The next wave of chips might be made of nanotransistors, which work on the molecular level. Ten million nanotransistors can fit on the head of a pin.

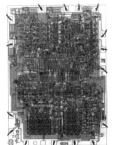  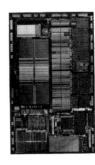 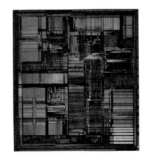 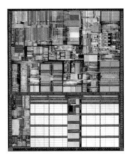 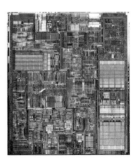

1971   1974   1985   1989   1993   1999   2000

USE SERVER FARM

The World Wide Web is free, isn't it? The servers that serve you—and that host web sites—operate all day, every day. Servers consume enormous quantities of electricity and require protection from human and environmental hazards. There is no free email.

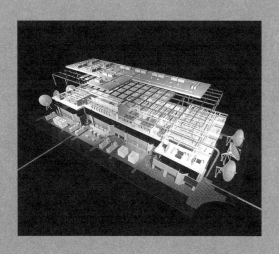

Metaphor

The all-pervasive nature of civil engineering works gives metaphorical power to civil engineering words. That is why information technologists often describe the Internet with civil engineering vocabulary.

During the second half of the 19th century, civil engineers in France coined the term *infrastructure* to describe railway networks. Recently, it has become *the* word in information technology.

Information superhighway was a common metaphor for the Internet. Unlike a paved highway, the Internet is largely a private place run by private industry.

Left: Server farm
Above: Microsystems advertisement, 2001

Server Farms

Once upon a time, companies operated their own servers for their own web sites. Keeping an increasing number of servers operational all day, every day, however, became a burden—and an opportunity for the entrepreneurs who offered to house the servers in a secure location.

Buildings called *server farms*, *Internet data centers*, and *mission critical facilities* house the telecommunications equipment and web servers that store web pages and create Internet connections. These buildings, linked to existing telephone grids and power lines, illustrate the Internet's rootedness in civil engineering infrastructure. This is the reality behind cyberspace. It is not placeless; it is rows of humming machines cooled by perpetually pumping air conditioners.

Right: Server farm
Far right, above: AboveNet building, seismically upgraded
Far right, below: Seismic isolation pendulum

Farm Construction

Server farms must be carefully built and maintained because of their power demands as well as the economic importance of their function. They must be able to support large loads in order to accommodate batteries and generators. They require high ceilings for fiber optic cables and racks of servers, and their reinforced roofs must be able to carry heavy cooling equipment. Server farms require special protection against fire and water; in California, they must survive earthquakes.

Earthquake Protection

An earthquake that disrupts Internet access could be economically devastating. Using seismic isolation, structures should be able to withstand earthquakes up to magnitude eight on the Richter scale.

The AboveNet building, an Internet service exchange facility in San Francisco, upgraded the seismic performance of its building with bearings placed between the building and its foundation. The bearings isolate the building from the earthquake's force.

Demand

In 2001, the demand for server space was growing at an estimated annual rate of 50%. This made it urgent to find a sustainable solution that would meet the energy demands of server farms. With some server farms earning an estimated $1200 a square foot, three times more than premier shopping malls, the economic imperatives are great.

Community

Communities often don't like server farms because of their industrial appearance. The warehouse-type buildings encourage little foot traffic because few workers are needed. The noise and air pollution from backup diesel generators, which are tested about once a week, raise health concerns. Frequently, though, server farms are housed in buildings that were vacant.

Server farms "are just as necessary as telephone poles and switching stations were in the last century. Just as people were spooked by electrical lights and telephones, they are spooked by these."

—Attorney TIM TOSTA, who represents server farm developers in San Francisco, 2001

Left: Servers and other equipment in racks

Right: Traffic jam during California blackout, May, 2001

Crisis

Some analysts believe that the California power crisis of the summer of 2001 is due to the massive electricity requirements of server farms and chip fabrication plants. In 2000, annual power consumption rose about 6% in California and 12% in the city of Santa Clara.

Some experts maintain that the Internet and the electricity it requires are merely convenient targets and that the real blame for the California energy crisis lies elsewhere.

Electricity use in watts per square foot	
HOME	1
COMMERCIAL OFFICE BUILDING	5
CHIP MANUFACTURERS	30-50
SERVER FARMS	75-100

Wealth and Power

Silicon Valley's high-tech companies are key to economic growth, but they burden an old energy supply infrastructure. Tech leaders bemoan the inadequacy of the state's power grid.

The consumption of electricity in Santa Clara County has grown by one-third since 1994. Santa Clara, a city in Silicon Valley, generates only 2% of its power within city limits. The Valley imports 90% of its power from elsewhere in California or out of state.

New Electricity, New Infrastructure

California gets its electricity from coal, natural gas, nuclear power, hydroelectric facilities, and wind. It is the largest consumer of electricity per capita in the world. About 80% of the state's generating plants are more than 35 years old. Besides several wind farms, little capacity has been built in the state during the last 15 years.

California imports 25% of its electricity from other states, many of which have weaker environmental laws. With utilities officials predicting 11% growth by 2004, utility companies are racing to build new generating plants. This will require renewal of the supply and distribution infrastructure.

"Conservation may be a sign of personal virtue, but it is not a sufficient basis for a sound, comprehensive energy policy."

—Vice President DICK CHENEY, 2001

Left: Transmission tower

Civil Engineers and Computers

Civil engineering was one of the first engineering fields to be computerized. In the late 1950s, businessmen and scientists were beginning to take advantage of computers. In the 1960s, in order to ease the labor and time required by complex calculations, civil engineers turned to the computer. Users had to write a program for each problem they wanted solved; this required both computing and engineering expertise.

A breakthrough occurred when Charles Miller, professor of civil engineering at MIT, developed a language called COGO (Coordinate Geometry System), which enabled engineers conducting topographical surveys to "speak" to the computer in engineering and geometrical terms. COGO, considered the first of many problem-oriented languages, became a success. It was later incorporated into Miller's Integrated Civil Engineering System, which brought computing to designers in transportation, urban development, and construction engineering.

In 1963, Steven Fenves and his colleagues at MIT developed a problem-oriented language named STRESS (Structural Engineering System Solver) for structural engineers.

In 1990, it was still unusual for each engineer to have a computer on his or her desk. An office would have a cluster shared by a group.

Have Computers Transformed Civil Engineering?

Plans are easier to modify. The speed of calculations that computers offer has made design more efficient. Greater precision is possible, as is virtual testing. As before, civil engineers tackle challenging and complex projects, design unusual and large-scale structures, and develop and test new concepts. Whether computerization results in superior and safer infrastructure design, construction, and management remains to be seen.

Computers are as good as the information they are given and the programs they run. They are no substitute for the training, knowledge, and intuition of the experienced engineer.

Seattle's Experience Music Project is covered with one of the most complex panel systems ever created for a building facade. No two of the 4,000 metal panels are alike. Its creators claim that the complexity of its design could not have been achieved without computer modeling.

Above: Charles Miller, former head of the Department of Civil Engineering at MIT, with student, 1960s

Below left: Engineers in the bridge department of Parsons Brinckerhoff Quade & Douglas, New York, 1957

Right: The Experience Music Project, Seattle, Washington

Below: The partners of Parsons Brinckerhoff Quade & Douglas review engineering drawings at the firm's New York headquarters, 1965

HOW MUCH DOES IT COST? 133

DISPOSAL
CATHODE-RAY TUBE

Trash

The writing is on the screen: between 2000 and 2007, up to 500 million personal computers will become obsolete. Analysts estimate that, in California alone, more than 6,000 computers are deemed obsolete each day.

Where will the darkened tubes end up? What is the cost of large-scale computer disposal? The answer is in the infrastructure.

> "Man's battle with nature has been won. Whether we like it or not, we are now burdened with the administration of the conquered territory."
>
> —Civil engineer OVE ARUP, 1970

Outmoded

Obsolescence is the dark side of Moore's law. As microchips become more powerful, the public rapidly perceives their computers as outdated. The average computer life span is now two to three years (it was four to six years in 1994), although it is believed that a microchip can function for much longer. The life span of a chip has never been established, and there is little market incentive for doing so.

The responsibility for disposal seems to fall on the consumer; but finding safe ways to dispose of computers is difficult, even in Silicon Valley. Many consumers are unaware that the moment they discard their hardware it begins a new life as hazardous waste.

Cathode-Ray Tube

Monitors are illuminated by cathode-ray tubes—vacuum tubes made of glass with electron guns at the far end. When a beam of electrons strikes the phosphor coated screen, it produces an image: a resumé, an email message, an on-line auction. Each monitor contains up to eight pounds of lead.

Dump

Garbage used to go to the dump. These open piles of garbage are rare today. Instead, trash is brought to an engineered landfill. Open dumps were dangerous; they housed vermin, were sites of uncontrolled fires, and allowed rainwater to carry bacteria and harmful chemicals (called "leachate") into groundwater. The U.S. Environmental Protection Agency banned dumps in 1979.

Landfill

The modern sanitary landfill contains three essential components: cover, liner, and leachate collection system. In a landfill, garbage is dumped, flattened, and at the end of the day blanketed by at least six inches of cover material. Layers of compacted garbage alternate with layers of cover material. Leachate is treated and discharged or recirculated in order to aid decomposition. The lining prevents ground and water contamination.

Landfills, developed in Great Britain in the 1920s and introduced in the United States in the 1930s, were formally mandated by the Resource Conservation and Recovery Act of 1992.

Lead

The EPA estimates that 80% of all discarded computers find their way into landfills. Landfilling a cathode-ray tube will cause it to be treated as regular trash, meaning it will be crushed and its lead will be released and potentially carried by rain to the groundwater. This process takes place over several decades.

Public Health

The engineered regulation of trash is essential for the maintenance of public health. The American Society of Civil Engineers recently asked its members to nominate civil engineering achievements that had the most positive effect on life in the 20th century: modern sanitary landfills made the top ten.

Background: Computer trash in landfill, Asia
Left: Sanitary Landfill, California

HOW MUCH DOES IT COST? 135

In the Closet

A 1991 study produced by H. Scott Matthews of Carnegie Mellon University predicted that, by 2005, 150 million computers would be cluttering the country's landfills. In 1997, Matthews changed his prediction to 55 million. This change was due in part to the use of recycled computers. Matthews believes, however, that a large number of computers have become "closetware." Individuals and companies, loathe to discard equipment, tend to stockpile it under desks and in closets.

United States government researchers estimate that three-quarters of all computers ever sold in the U.S. are closetware. Were all these tossed out at once, the nation would face a crisis with major budgetary and environmental ramifications.

Out of the Closet

The infrastructure for convenient, safe, and inexpensive computer disposal does not yet exist.

The numbers of computers being sent to landfills remains potentially dangerous. Forty percent of all the lead in American landfills was once part of a television or a computer monitor. Two to five percent of municipal solid waste consists of electronics. This percentage is rising more rapidly than those of other kinds of trash.

Cost of Disposal

Cathode-ray tubes are not allowed in municipal landfills in California. Instead, they must be brought to a recycler or a hazardous waste facility. Reports circulate, however, of municipal landfills accepting the lead-filled tubes.

The cost to taxpayers and local governments for collection and proper handling of cathode-ray tubes could exceed $1 billion over the next five years.

"In the end, garbage is basically just a lack of imagination."

—Architect MINDY LEHRMAN CAMERON, 1993

Closetware

Corporate Responsibility

The European Union is developing a plan that would make computer firms responsible for the life-cycle environmental impact of their products, from the extraction of raw materials to the disposal of the completed product.

Some companies in the United States are instituting sustainable practices on their own. Compaq, for instance, takes 200,000 computers a year from corporations in North America. IBM introduced the first computer that uses 100% recycled resin in all major plastic parts in 1998, and also has a buyback program. Several manufacturers phased out glue adhesives, thereby making parts easier to separate during recycling; replaced nuts, screws, and bolts with snap-to-fit parts; reduced the number of plastics used; and labeled those plastics to ease the job of recyclers.

Hewlett-Packard has its own recycling facility. It recycles 3.5 million pounds of electronic equipment a month, but only accepts machines from companies, not individuals.

Waste Abroad

Smelting of metal from computer parts happens in places far from population centers such as the San Francisco Bay Area. It is usually cheaper to export metals for smelting to countries that have less restrictive pollution laws. In 1997, recyclers shipped abroad approximately one million of the 1.7 million monitors that they handled. Meanwhile, the millions of monitors sitting on and under desks and in closets are running up a bill that Americans will have to pay.

Above: Two recyclers remove a motherboard from a computer

Center: Granulated computer equipment coming off a vibrating screen that separates copper, gold, and silver

Below: Cathode-ray tubes separated from the waste stream

HOW LONG WILL IT LAST?

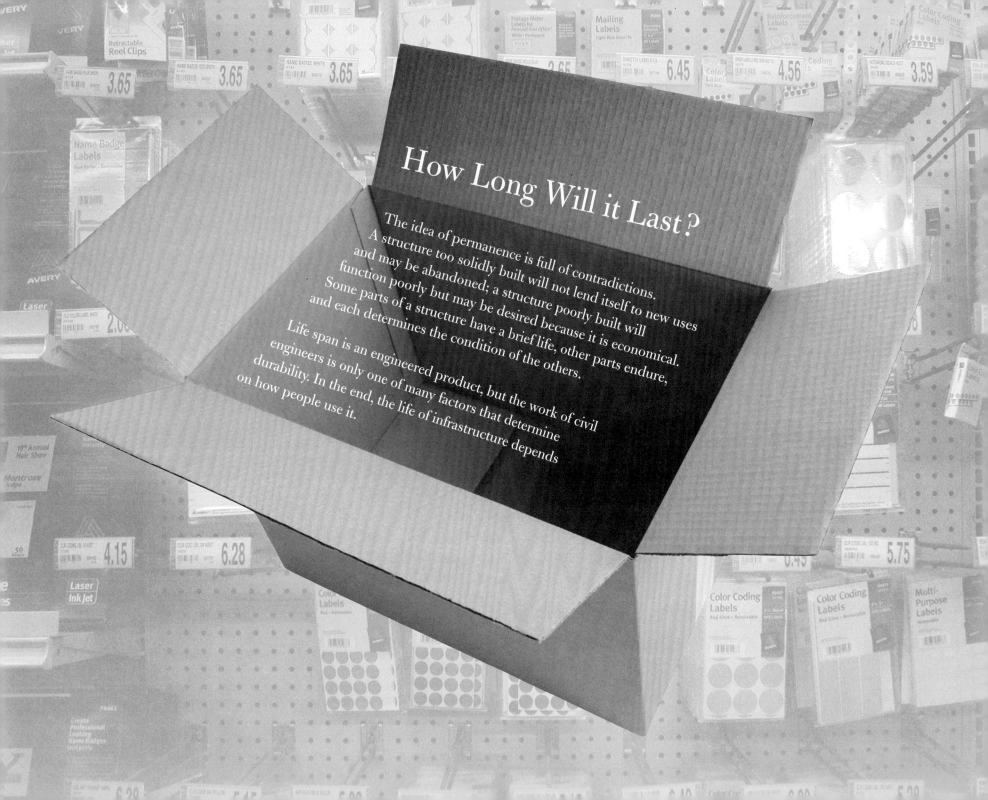

How Long Will it Last?

The idea of permanence is full of contradictions. A structure too solidly built will not lend itself to new uses and may be abandoned; a structure poorly built will function poorly but may be desired because it is economical. Some parts of a structure have a brief life, other parts endure, and each determines the condition of the others.

Life span is an engineered product, but the work of civil engineers is only one of many factors that determine durability. In the end, the life of infrastructure depends on how people use it.

PERMANENT

Permanence is an illusion built on foresight, commitment, expertise, and access to resources.

PERMANENT
- 142 Adaptability
- 144 Maintenance
- 146 Sustainability
- 147 Path Dependency

IMPERMANENT
- 148 Physical Life / Economic Life
- 150 Temporary
- 152 Doomed
- 154 Failure

GONE
- 156 Knowledge
- 157 Neglect

FOREVER
- 158 Eternity
- 160 Memory

Long-lasting infrastructure must be adaptable because societies change more rapidly than their ability to change buildings and sewer systems. "Durable" structures have been maintained, repaired, and renovated—if not rebuilt. To be effective, regular care must be coupled with knowledge and experience.

Understanding permanence means understanding the past in order to shape the future.

ADAPTABILITY | Big Box Store

Who would have imagined that large numbers of Americans would want to drive trucks? Or that they would shop in stores so big they could comfortably drive their SUVs inside?

Infrastructure must accommodate social change. If it cannot, a new use must be found, or it will be abandoned.

A tidal wave of skyscraper nostalgia hides the fact that, in the United States, the age of the high-rise ended long ago. Wow! The Sears Tower is big—but in 1992 Sears moved 5,000 employees 35 miles, from its downtown Chicago skyscraper to the suburbs. Low-lying big boxes and ground-hugging corporate headquarters are the American icons at the beginning of the 21st century.

The big box store has become a symbol that some people love to hate because of its appearance and its impact on small businesses, community life, and the environment. Yet the big box represents what Americans seem to want: low prices, disposable products, and a lifestyle that revolves around automobiles and trucks. How long will these stores—and this lifestyle—last?

Big box stores, U.S.A.

Big Box Construction

The big box is both store and billboard. Designed for vehicular access and visibility, it ranges from 25,000 to 200,000 square feet—the size of four football fields.

Architects and civil engineers develop a set of standards for "roll-out programs," that include the people, materials, and methods that allow retail chains to build quickly and economically. Big box store structures typically are supported by eight and ten-inch rectangular steel tubes; lightweight beams support the roof. Exterior walls consist of CMU (concrete masonry units, once known as "cinder block") and EIFS (exterior insulation and finish systems). These layered surfaces include insulation board, a water-resistant material, and a colored stucco-like top coat. The insulation board can be cut into various shapes, including classical columns and arched openings that lend an air of permanence.

Big boxes are designed for "demand shoppers:" people who know what they want and what they want to pay. In order to remain competitive, discount retailers must keep building costs down and keep up with the changing taste. Remodeling is anticipated every five to seven years. Anticipated building life span is ten or fifteen years. In comparison, a suburban office building is estimated to have a 30-year life, about the same as predicted for skyscrapers circa 1930. Many skyscrapers built at that time, however, are still standing.

Big Box Abandoned

When a traditional retail store went out of business, a new entrepreneur could come claim the space. What will become of vacated big boxes and parking lots?

Abandoned big box store, Loudoun County, Virginia

Below: Big box exterior wall finish and concrete masonry units; *Below right*: Big box store under construction

MAINTENANCE

Infrastructure appears strong, but it is fragile. Even the simplest structure must accommodate a network of users, abusers, materials, and environments—all of which change and change again.

When a building, road, or drainage pipe opens for use, construction is not over. It is the beginning of *maintenance*, the phase of construction that lasts the longest. Without upkeep, the life of the structure and the lives of its users are in danger.

Paint

Thousands of pounds of "International Orange" protect San Francisco's Golden Gate Bridge (1937) from the airborne salt that rusts steel. The color, suggested by architect Irving Morrow, advertises infrastructure that is well-maintained. For almost 30 years after its opening, painters provided touch-ups. In 1968, the effects of corrosion led civil engineers to replace the original lead coatings. In 1990, painters applied an acrylic topcoat to meet air quality requirements.

The maintenance crew of the Golden Gate Bridge does not consist solely of its 38 painters, however. Seventeen ironworkers also hang high above the San Francisco Bay to replace rusting rivets and steel. They temporarily remove parts of the bridge to allow the painters access to out-of-reach areas.

Following several storms that rocked the Golden Gate Bridge, and the collapse of the Tacoma Narrows Bridge, engineers added 500 tons of steel bracing. Currently, a $297 million seismic retrofit is underway to protect the bridge during an earthquake. Were the Golden Gate to fail, the replacement cost is estimated at $1.4 billion, not including disruption to the economic and social life of the region.

Facing Page: Painter atop the 746-foot south tower of Golden Gate Bridge
Below: Golden Gate Bridge, San Francisco

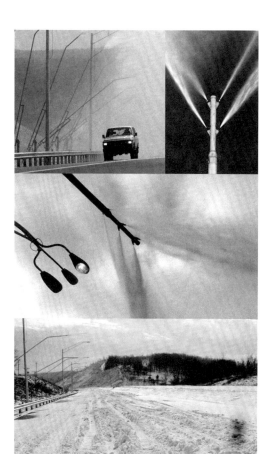

Above: Experiments in progress on the Smart Road
Below: Smart Bridge, Blacksburg, Virginia

Research

Designing infrastructure for maintenance is as crucial as maintenance itself.

The Virginia Department of Transportation and Virginia Tech's Center for Transportation Research opened a $35 million Smart Road in Blacksburg, Virginia. The work that civil engineers carry out there will help shape the future of road design and maintenance.

On the Smart Road, Virginia Tech will test research vehicles equipped with safety-enhancing intelligent transportation system technology such as traction control, and night and bad-weather vision response. Civil engineers with the Virginia Department of Transportation will study the road in order to develop designs that interact with new technologies, and to create proactive strategies for highway maintenance.

The Smart Road contains a mix of materials and research tools. More than 600 sensors generate data on pavement performance. The lighting includes 90% of the existing types of roadway illumination. A 500,000-gallon water tank feeds 40-foot tall rotating snow towers that create fog, ice, snow, and rain, in order to assist the study of roadway dangers. When the Smart Road is extended to connect Blacksburg to the interstate highway, two lanes will remain a research laboratory.

SUSTAINABILITY

Plastics play an increasingly large role in construction. Plastic packaging is instant trash. Where do old tires roll? Everything has to end up somewhere. The absence of water, air, and sunlight in landfills means that most garbage will be preserved for future archeologists curious about American life today.

In order to protect the places Americans live, innovators including civil engineers try to find ways to recycle, safely store, or reduce the amount of garbage-in-the-making.

Plastic

The American lifestyle is increasingly based on plastic. This lightweight and strong material is not as temporary or safe as it appears. Plastic products have a service life ranging from a few weeks to 40 years—but they are not biodegradable and options for disposal are limited. If recycled—a big if—those famously unnecessary CD cases can become non-recyclable egg cartons. If recycled, CDs might be reborn as automotive parts. Fifteen percent of discarded plastic is incinerated; the rest is sent to landfills, where it takes up approximately 20% of the volume. According to the University of British Columbia, plastic six-pack holder rings will take 450 years to biodegrade and aluminum cans will take 100 years. Plastic bottles will "never" biodegrade.

Pipe made of polyvinyl chloride (PVC), a particularly durable plastic, is comprised of 56% chlorine, in addition to other hazardous additives. This popular building material is piling up in many countries. PVC presents a particular challenge to waste disposal technologists; it cannot be incinerated or dumped because of its hazardous contents.

Tires

Tires are forever—or so it seems. They are a symbol of the challenge presented by "products" that rapidly become "garbage." The 260 million tires Americans discard each year pose fire, health, and environmental hazards.

Tires are everywhere. For example, the 2.9 million citizens of Iowa discard three million tires each year. Thirty-four million Californians discard thirty-one million tires and three million more are stockpiled in the state. In 1977, the *Charleston Gazette* stated that the six million discarded tires in West Virginia would allow citizens of the state "to have three tire swings in their backyard, with 600,000 spares."

Sixty-six percent of discarded tires are recycled today versus 11% in 1990, according to the Department of Energy. Civil engineers reuse tires in levies, septic tank systems, and as a substitute for gravel. Old tires are used as fuel and transformed into playgrounds and welcome mats. A civil engineer in Phoenix developed asphalt that includes tire rubber.

PATH DEPENDENCY

Civil engineers, elected officials, manufacturers, contractors, real-estate developers, and many others—including you—make decisions. The result is a curb cut, a sewer system, an airport, a winding street with suburban homes. Infrastructure creates a path or pattern that lasts, while society changes.

The physical and spatial pattern established by infrastructure is strengthened by vast economic, social, and emotional investments. This means that paths endure.

Railroad tracks, Meridian, Mississippi

Street

Residents of New Amsterdam—founded in 1624—established the layout of the streets below Wall Street, the northern boundary of the settlement until the late 17th century. The street pattern, first shown in map form in 1660, is typical of a Dutch town of the middle ages.

In 1811, New York established a grid plan for its streets. At that time, most individuals did not own horse-drawn carriages. There were no automobiles. Buildings were two or three stories tall. It was believed that the heaviest traffic would run east-west.

In the 21st century, two historic street patterns are the basis of New York City life.

Lower Manhattan, 1660

Right-of-Way

America's aging rail system has a new neighbor running alongside its tracks: the telecom industry's fiber optic cables. Like railroads, fiber optic cables need to be near both established population centers and areas of growth.

Railroad tracks make good paths for telecommunications cables because they offer straight, cleared routes. Were these paths to be assembled from scratch, the cost would be enormous. The rail-cable partnership also works well for railroad companies that collect money from the use of their right-of-way. These companies have themselves spawned telecommunications enterprises such as Sprint (Southern Pacific Railroad International Telecommunications). From 1980 through 2000, communications companies installed 85 million miles of fiber optic cables in the United States—33 million in 2000 alone.

IMPERMANENT

PHYSICAL LIFE / ECONOMIC LIFE

Some structures are meant to last only as long as the activities they house are deemed profitable. Some are meant to be assembled and disassembled. Some are condemned by politics or public taste. Others fail through error or ignorance.

"The modern trend... is to erect buildings not for perpetuity, as in ancient days, but for investment, with a view to replacement when their estimated usefulness is over."

—*New York Times*, 1926

How much is an owner willing to invest in the original construction? How much money will be devoted to maintenance? What is the relationship between popular taste and the longevity of a building?

When the cost of using a structure is deemed higher than its worth, its life may be over.

Right: Demolition of Kingdome

Below: Kingdome, Seattle, Washington

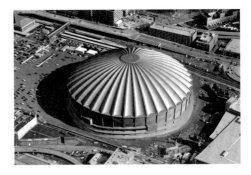

Stadium

Dubbed "the mushroom," "the concrete cupcake," and even "the ugliest man-made structure ever built," the Kingdome Stadium, home to Seattle's baseball and football teams, was imploded to the sound of cheering crowds on March 26, 2000.

The stadium was considered an engineering marvel when it was completed in 1976, and with its 10,500 tons of steel reinforcement, it was supposed to last for many seasons. But sports fans and players found that the Kingdome could not satisfactorily accommodate both baseball and football. Its appearance earned it no defenders. Hours before a game in 1994, several tiles fell from the concrete dome, the world's largest, into the stands. The resources and the desire to preserve the structure did not materialize.

A dome was once considered a necessity in rainy Seattle. Although the new baseball stadium has a retractable roof, the new $430 million football and soccer stadium is uncovered. King County still owes money on the first structure, one-third of whose rubble was planned for use in the new construction.

Washington State Football/Soccer Stadium and Exhibition Center (*foreground*); SAFECO Field (*background*)

Paving

Petroleum asphalt covers at least 90% of the roads in the United States. When designing the surfaces of these roads, civil engineers face what historian Bruce Seely calls "a classic engineering challenge:" they must balance economy and durability, in order to provide a safe ride in a wide range of climates and conditions.

Pavement must perform in an unforgiving environment, where soil, foundations, paving materials, and ever more numerous and heavier cars and trucks come together at high speed. How long should asphalt last? Some engineers use traditional methods and materials, while others advocate state-of-the-art. At one time, federal regulations seemed to require roads to fail before money for maintenance would be provided.

The Federal Highway Administration estimates that 58% of the nation's urban highways are in poor, mediocre, or fair condition. The American Society of Civil Engineers' *Report Card for America's Infrastructure* gives U.S. roads a rating of D+.

HOW LONG WILL IT LAST?

TEMPORARY World's Fair

Structures intended to have a short life encourage innovation because many standard requirements do not apply. They can be erected in remote areas and in places without the infrastructure that supports permanent buildings. Because temporary structures allow for rapid change, reuse, and recycling, they present an opportunity for rethinking what buildings are and how they should function.

> "Every problem is a challenge and many problems were met in the building of the Fair that were never before encountered. The engineers have given a hundred new and valid answers to a hundred major and perplexing questions. They have created things which many said were fantastic, and others, impossible. And that is because nothing can defeat a combination of experience and courage."
>
> —Official Guide Book, New York World's Fair, 1939

In the years between the Great Depression and America's involvement in World War II, the New York World's Fair Corporation planned what its president, Grover Whalen, called "the greatest civil engineering feat of the century:" the 1939-1940 New York World's Fair. The theme, "Building the World of Tomorrow," highlighted an optimism that was intended to lift New York and America from economic and psychological doldrums. The symbol of the fair was the Trylon and Perisphere, a 610 foot tower and 180-foot sphere, framed in steel and covered with stucco.

Robert Moses, the force behind the construction of much of New York's infrastructure, expected the fair to transform a former trash dump in the Flushing Meadows section of Queens into a park equal to Manhattan's Central Park. The fair also gave Moses a rationale for expanding New York's highway infrastructure.

The fair's catalogue likened construction of the fairgrounds to building a new city of 800,000 people on 1200 acres of uninhabited land. The underground pipes and conduits were enough to serve a city of 1,000,000. While the fair attracted about 45 million visitors, it didn't make a profit, and Flushing Meadows never became another Central Park. Steel from the fair's structures was sold or used for the war effort—its contribution to the world of tomorrow.

Below: "Streets of the World" and "World's Largest Cash Register," New York World's Fair, 1939

Facing page: Trylon and Perisphere under construction

"It sounds imposing to say, 'We are building for all time.' It might be much better business to say, 'We are building for fifteen years.' The canvas tent of the traveling circus, the plaster buildings of a World's Fair, the granite and marble of a municipal building, differing as they do, yet each exactly meets the requirements of the particular case."

—GEORGE HILL, "The Economy of the Office Building," *Architectural Record*, 1904

Concert Stage

Designers of rock concert stages provide a framework that must be easily transported, erected, and dismantled, in some cases up to 200 times a year. These temporary stages support complex entertainments, often in areas far from centers of population.

In U2's *Popmart* world tour (1997-1998), architect Mark Fisher worked with engineers to create a 100-foot tall golden arch that paid homage to a fast food chain while providing support for the sound system and light fixtures. The set also featured the largest video screen (165 by 50 feet) created up to that time.

Pink Floyd's *Division Bell* tour (1994) featured a 130-foot wide semi-circular arch made from trusses that slid into one another, could be assembled in seven hours, and were dimensioned to fit exactly across the width of a truck for travel. A water-filled base created the foundation. Two towers supported the arch and carried the sound system and enclosures, from which inflatable pigs sprang.

Pink Floyd, *Division Bell* World Tour, 1994

DOOMED | Beach

No matter what is done, and despite the best efforts of civil engineers, some structures cannot survive. Once, the prevalent attitude was that nature had to be "conquered" no matter what the cost in dollars and human hardship. In recent years, attitudes have begun to change, although the battle continues and, on occasion, victory is declared—for a while.

Every year, hurricanes and tides erode North Carolina's beaches, causing substantial property damage. Millions of dollars have been spent on repair. Should construction be permitted in a landscape subject to recurring "natural" disaster?

Civil engineers stand on both sides of the issue. Some are involved in efforts to rebuild beaches with sand dredged from the ocean floor. Others point to the need to halt the construction of housing and shopping centers in coastal zones; these structures halt the flow of sediment, preventing natural replenishment of the shore.

Below: Cape Hatteras lighthouse, before its move, 1999; *Right*: Cape Hatteras Lighthouse, moved 2,900 feet to protect it from the ocean *Below right*: Condominium resident, North Carolina, 2001

Roof

Hurricane Andrew proved one of the most costly natural disasters in U.S. history when it hit Florida in 1992, causing $20 billion worth of damage. How "natural" was the destruction of thousands of homes? Civil engineering research conducted afterwards showed that improperly driven nails resulted in many roof blow-offs.

In order to insure that new wood-framed houses withstand hurricanes, the new code requires the use of metal connections capable of holding the frame together, from foundation to roof. Previously, it took a lifting force of about 37 pounds to separate a 2x8" roof beam nailed to the top beam of a wall. The new 1/16-inch-thick steel strap nailed over the same joint requires a force of about 300 pounds.

Why were so many structures vulnerable to hurricanes in an area with a long history of storms? It was a result of what historian Ted Steinberg calls the "federalization of risk," in which the federal government permits and subsidizes the construction and rebuilding of residential infrastructure and housing in disaster-prone areas.

Below: Owner with destroyed home; *Right*: Aftermath of Hurricane Andrew

FAILURE | Bridge

Why do structures fail? It might be an engineer's belief in the safety of a design, an unexpected act of nature, political conditions that encourage specific ways of building, poor workmanship, lack of construction supervision, acts of war, attempts to save money, or calculation errors. Failure can occur due to conditions nobody could have known or predicted. In all cases, engineering failures are significant because they are learning experiences.

"There is no finite checklist of rules or questions that an engineer can apply and answer in order to declare that a design is perfect and absolutely safe, for such finality is incompatible with the whole process, practice, and achievement of engineering."

—Engineer and historian HENRY PETROSKI, *Design Paradigms*, 1994

On December 15, 1967, the Silver Bridge (also known as Point Pleasant Bridge) fell into the Ohio River, killing 46 people. Why did this happen? After four years of study, engineering investigators from the National Transportation Safety Board determined that one steel eyebar contained a microscopic flaw that resulted from the manufacturing process. The bridge design was unusual because it had no redundancy. In other words, the design enabled the failure of one part to cause the entire bridge to fail.

One of only three eyebar-chain suspension bridges of this type ever built, the Silver Bridge was constructed in 1927 to link Point Pleasant, West Virginia and Kanauga, Ohio. The bridge relied on steel suspension bars linked with a steel pin through the eyelet at each end. The bars resembled a bicycle chain.

As a result of the failure, the federal government responded by establishing the National Bridge Inspection Standards under the Federal-Aid Highway Act of 1968. Before this disaster, little support existed for a federal bridge inspection program. Congressional hearings following the collapse demonstrated that few states knew how many bridges they owned, and many did not have formalized inspection or record-keeping procedures.

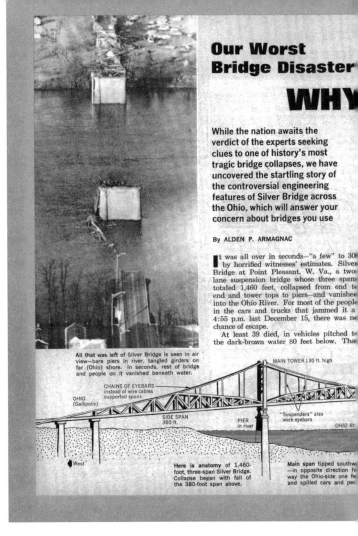

Silver Bridge, *Popular Science* (1968)

The second eyebar-chain suspension bridge in the United States was replaced. The third, in Brazil, had four rows of eyebars on each side, compared to the American bridges' two. It was deemed to have sufficient redundancy and continued in service until recently. The new Silver Memorial Bridge was built further south in 1969.

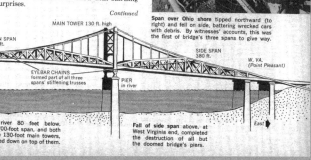

Dam

Civil engineer William Mulholland, who participated in the design and construction of the Hoover Dam and Panama Canal, and was in charge of the construction of the Los Angeles (Owens Valley) Aqueduct, oversaw the building of the St. Francis Dam (1924-26). Its purpose was to provide Los Angeles with an auxiliary water source. The dam held 12.5 billion gallons —enough to supply Los Angeles with water for one year.

Minutes before midnight on March 12, 1928, the St. Francis dam collapsed. Water swept 50 miles, from the Santa Clara Valley towards the Pacific Ocean, killing almost 500 people, destroying thousands of acres of land, as well as railroad lines, livestock, and bridges.

The investigation that followed was inconclusive. Rumor circulated about the possibility of sabotage by disgruntled farmers and ranchers. Recently, engineering geologist J. David Rogers suggested a number of possible factors, including the fact that the eastern edge of the dam was built on an ancient landslide—a geological factor that Mulholland had no way of recognizing. None of this, however, necessarily caused the collapse. Rogers believes that the decision to heighten the dam, so that it could hold additional water, led to the stress that caused the collapse.

The tragedy led to the establishment of a dam safety agency and new engineering testing criteria.

Below left: Silver Bridge; *Below*: St. Francis Dam, 1928

Remains of dam, sometimes called "the tombstone"

GONE

The house or apartment you live in, the roads you drive on—these will one day be gone.

Through neglect or destruction, all civil engineering works will eventually disappear.

KNOWLEDGE

Can anything positive come out of the horrific terrorist attacks of September 11, 2001?

No structure has been designed to survive the impact of a fuel-laden Boeing 767 jet—an impact equal to 1/25 of the atomic bomb that leveled Hiroshima. Moreover, it is unlikely that anyone would want to live or work in a massive windowless "secure" hulk —should a developer be willing to pay for its construction.

Precisely how the towers of the World Trade Center collapsed is a question under investigation by a team of engineers established by the American Society of Civil Engineers. Such teams have studied disasters including earthquakes in El Salvador, India, Peru, and the United States. Engineers will use the knowledge gleaned from the wreckage of the World Trade Center to safeguard the lives of people in high-rise buildings throughout the world.

World Trade Center site, January, 2002

NEGLECT | Stations

"Outdated" or "unwanted," are terms used to describe infrastructure that is likely to be neglected. Deliberate disregard includes a lack of planning, inadequate maintenance, or a withdrawal of funding. Structures and infrastructures don't "become" obsolete, they are made obsolete by choice.

In 1920, most Americans traveled between cities on a transportation infrastructure whose construction was synonymous with the construction of the United States: the railroad. There were more than 80,000 depots and railroad terminals, and more than 50,000 passenger cars; 1.8 million people—one in every 23 Americans—worked in the rail system. By 1930, the Depression and the rise of the automobile caused a plunge in ridership.

After World War II, the federal government's generous support of highways, airports, and suburban expansion did not extend to passenger rail. Between 1946 and 1953, passenger numbers decreased fivefold and railroad companies abandoned one-third of their tracks. Abandoned stations, some restored and used as homes or offices, stand as witnesses to the time when mass transit defined American life.

Train station, Tuscaloosa, Alabama, 1993

FOREVER

Everyone likes to believe that some things will last forever, unchanged. The only thing that goes on forever is change.

ETERNITY

Engineers design some structures to last "forever." Some structures last so long that they come to symbolize eternity.

Nuclear Waste

The $58 billion federal nuclear waste repository under Yucca Mountain in Death Valley, Nevada, if permitted to open, will include a 100-mile network of tunnels designed to hold uranium fuel pellets. Scientists believe that they decay to safe levels in 10,000 years. Nothing built has ever lasted that long. The Pyramids at Giza are 4,000 years old.

If waste storage begins as planned in 2010, the regulatory period will end in the year 12010.

Top: Protesting against proposed Yucca Mountain nuclear waste repository, 2000

Bottom: Workers inside Yucca Mountain tunnel

Yucca Mountain, Nevada

Cathedral

With 50,000 visitors a day, the Cathedral of Notre Dame of Paris is probably the most visited Gothic cathedral in the world. Its site, in the center of the city, has been considered sacred since pre-Roman times. Skilled builders participated in the construction of the cathedral between 1163 and 1345.

The 800-year history of Notre Dame, built in stages over two centuries, is a story of change. During the French Revolution, many of Notre Dame's treasures were destroyed or stolen. Subsequently, it was used for activities including food storage. During the Paris Commune of 1871, it was almost burned.

Architect Viollet-le-Duc oversaw a massive restoration program beginning in 1845; he rebuilt the spire and created new gargoyles. During the 1990s, another restoration brought the cathedral to an imagined Gothic glory. Today, Notre Dame continues to draw tourists, for whom the seemingly solid structure is a symbol of eternity.

Notre Dame, Paris, France, 1975

"Assuredly the Cathedral of Notre Dame of Paris is, to this day, a majestic and sublime edifice. But noble as it has remained while growing old, one cannot but regret, cannot but feel indignant at the innumerable degradations and mutilations inflicted on the venerable pile, both by the action of time and the hand of man."

—VICTOR HUGO, *Notre Dame de Paris* (1831)

MEMORY | Documentation

The future of civil engineering is inseperable from the experience and knowledge embedded in its history. Where can we find the infrastructure of the past? Printed and digital records prepare infrastructure for its final incarnation, in the mind.

The American Society of Civil Engineers, the National Park Service, and the Library of Congress founded the Historic American Engineering Record (HAER) in 1969, in order to document the disappearing engineering and industrial structures and sites in the United States. It was established as a complement to the Historic American Buildings Survey (HABS), founded in 1933.

College students who study engineering, history, and architecture help prepare most of the photographs, drawings, and histories of structures that would otherwise vanish from memory. The HABS/HAER collections, which include 68,000 photographs, are among the most used in the Prints and Photographs Division of the Library of Congress.

Sands Point Lighthouse, North Hempstead, New York, built 1806–9

Port Jervis Roundhouse, Port Jervis, New York, built c. 1848

Waco Suspension Bridge, Waco Texas, built 1870; altered 1914

Publication

Where are the civil engineering structures of yesteryear? Take bridges: hundreds of thousands built during the 19th century are gone; the paperwork behind their design and construction is gone; the engineers and workmen who built them are gone. Some of these bridges, and the debates behind their design and construction, survive only in fragile books and journals. These publications enable historians to uncover the basis of industrial society. Through text and image, forgotten civil engineers and the origins of contemporary technology may re-enter memory, and live again.

Wilhelm Friedrich Kuhn, *Theoretisch-praktisches Handbuch des Straßen- Wasser- Brücken- and Hochbauwesens für Anfänger in der Baukunst.* Ulm: By the Author, 1832.

Michel Chevalier. *Histoire et description des voies de communication aux Etats-Unis et des travaux d'art qui en dependent.* Paris: Charles Gosselin, 1840-41.

Howard Douglas. *An Essay on the Principles and Construction of Military Bridges, and the Passage of Rivers in Military Operations.* 2d ed. London: Thomas and William Boone, 1832.

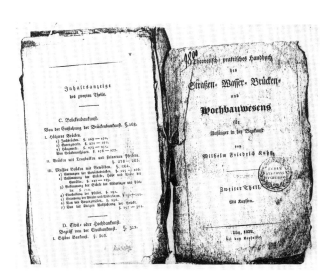
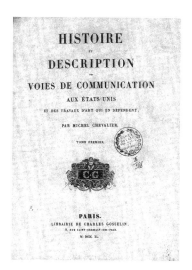
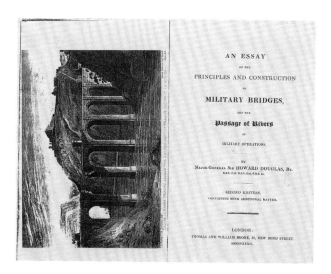

Loudoun County, Virginia

Afterword

How Long Will the Future Last?

Tomorrow does not just happen. The future is built right on top of infrastructure. Safe water and air? Secure homes? New businesses? Telephone calls? Parks and parking lots? Transportation to the other side of town— or the world? The answers make today and tomorrow possible.

As pedestrian or driver, coffee drinker or dishwasher, word processor or number cruncher, impulse buyer or bargain hunter—and as voter— you help determine how long the future will last. Civil engineers serve you by designing the future you choose. So go ahead! Walk across the street. Drink a cup of coffee. Download that file. Go shopping. Drive home.

Life: it's infrastructure and you.

Acknowledgements

We are very grateful to the people who helped us create *Me, Myself and Infrastructure*.

The project was well advanced when the United States was attacked on September 11, 2001. This event strengthened our resolve to complete *Me, Myself and Infrastructure* unaltered and with minimum delay. We would like to thank Karen Kitchen and Melissa Coley for believing in the potential of this project at an early stage. We truly regret that the exhibition did not appear, as planned, within the World Financial Center's Arts & Events Program.

We'd like to thank the New-York Historical Society for becoming our New York host and collaborator. Kenneth Jackson, Jan Ramirez, Rob Del Bagno, Kathleen Hulser, Jennifer Jensen, Holly Hinman, Nicole Wells, Marybeth Kavanagh, and Suzanne Kammin are among N-YHS staff who have assisted us.

At the Science, Industry, and Business Library of the New York Public Library, we'd like to thank Kristin McDonough, John Ganly, and their staff.

We have received invaluable assistance from many people, including: Mike Beavin, Aaron Byrd, Howard Decker, Judy DeMoisy, Susan Dennis, Ted Dowey, Brad Fagrell, Donald Friedman, Adam Hart-Davis, Patrick Howey, Steve Huey, Bea Hunt, Heather Kilpatrick, Frank Lombardi, Steven Lubar, Charles Maikish, Ann Miller, Glenn Orenstein, Peter Rothenberg, Paul Rudisill, Charles Scawthorn, Zach Schrag, Bruce Seely, Dan Spurling, Jeffery Stine, Kara Swisher, Selma Thomas, Karen Turner, Jon Wilkman, and Jeffrey M. Zupan. Special thanks to Rae Zimmerman and Nate Gilbertson at the Institute for Civil Infrastructure Systems at New York University; Priscilla P. Nelson and her staff at the National Science Foundation; and Lorraine Whitman of the Salvadori Center.

We are indebted to those who discussed with us questions relating to pedestrian safety and transportation issues: Michael Ronkin of Oregon DOT and Courtney Duke in the Portland Office of Transportation Planning; and Michael Fishman and Sam Schwartz of the Sam Schwartz Company; Thomas W. Brahms of the Institute of Transportation Engineers; Robin Overstreet and URS Corporation in Atlanta; Rick Stevens of the Washington Metro; the Maryland Governor's Office of Smart Growth; and Randy Wade at New York DOT. We'd like to thank Victor Ross for generously sharing his experience and his materials. We are grateful to the Virginia DOT/Springfield Interchange team, including Larry Cloyed and Steve Titunik.

For images of Silicon Valley, we are indebted to Carol McCarthy, Pam Morrison, and Mary Hanel of the City of Santa Clara; Carol Fisher and Michael Lapointe of the Mountain View Historical Association; and Meryl Ginsberg of Varian Medical Systems. Paula Jabloner and Melissa Johnson of History San Jose located artifacts for us. For sharing expertise and artifacts we are indebted to Paulina Borsook and Jim Fisher; Alex Twining and Anne Katzen of MetroNexus; and Jeffrey Konspore, Edward Campbell, and the team at Per Scholas.

We could never adequately thank the researchers and writers listed on the title page of this book. Tiffany Anderson and Lauren DeMille know that this project would not exist without their extraordinary efforts. Mona Dreicer, Emily, Adrian, and Arielle Tylim, Ann Ostrander and Michel St. Pierre, Tracy Coffing and Ron Rogos, and Mary Mahon provided us with resources and opportunities that contributed immeasurably to our work. Neil Budzinski provided imaginative answers to our research requests. Thanks also to Tami Gatta, Josephine Yeh, photographers David Arcos, Brandon Fernandez and Jim Sichinolfi, Mark Hado and George Brosnan of Copytone Reprographics, and to Suzanne Salinetti of the Studley Press.

We offer extra-special thanks to our dedicated and selfless video production team: Anna Strout, Amy Wood, and Lynn True.

Most importantly, we would like to thank the American Society of Civil Engineers, in particular the members of the Steering Committee of the ASCE 150th Anniversary, who backed this unusual approach to their profession. ASCE Foundation staff was supportive and patient: Curtis Deane, Paul Skoglund, Casey Dinges, Gretchen Galuska, Rebecca Walters-Parrish, Chris Weber, John Durrant, Norida Torriente, and Jack Bruggeman. Jane Howell's early support for this project was crucial. We collected information from engineers and their colleagues throughout the United States—thanks for taking the time and thanks for the infrastructure.

Bibliography

Most publications that deal with the contemporary concept of "infrastructure" are intended for a professional audience. Those aimed at a popular audience are out of date. The *Me, Myself and Infrastructure* bibliography that follows will enable you to begin to satisfy an interest in infrastructure.

An up-to-date starting point is the web site of the American Society of Civil Engineers: www.asce.org. ASCE and the American Public Works Association both provide a multitude of new publications, but none have been as inclusive as two older works, written from the civil engineers' viewpoint: "Centennial Transactions" of *ASCE Transactions* (Volume CT, 1953) and the APWA's *History of Public Works in the United States, 1776-1976* (1976). Material can be found in collected conference papers, such as *Perspectives on Urban Infrastructure* (1984), edited by Royce Hanson; *Infrastructure: Maintenance and Repair of Public Works* (1984), edited by Alan H. Molof and Carl J. Turkstra for the New York Academy of Sciences; "A New Millennium Colloquium on the Future of Civil and Environmental Engineering" (2000), produced by the MIT Department of Civil and Environmental Engineering (web.mit.edu/civenv/www/colloquium.html); and the "First Annual Conference on Infrastructure Priorities" (2001), produced by the Institute for Civil Infrastructure Systems at New York University (www.nyu.edu/icis/InfraPriorities). *Infrastructure and Urban Growth in the Nineteenth Century* (1985), published by the Public Works Historical Society and *Technology and the Rise of the Networked City in Europe and America* (1988), edited by Joel A. Tarr and Gabriel Dupuy, provide historical perspectives.

Every once in a long while, a civil engineer such as Georg C. Mehrtens, William Barclay Parsons, or James Kip Finch begins to write history in order to explain the profession to itself and to a broad audience. Today, the public has ready access to the numerous thoughtful works of Henry Petroski. In recent decades, however, no scholar has produced a comprehensive history of civil engineering. Daniel Calhoun's *American Civil Engineer: Origins and Conflict* (1960) and Edwin Layton's *Revolt of the Engineers* (1971) are available in libraries. Historians and sociologists study the civil engineering profession; see, for example, Peter Lundgreen's work, "Engineering Education in Europe and the U.S.A., 1750-1930: The Rise to Dominance of School Culture and the Engineering Professions," in *Annals of Science* 47 (1990): 33-75. ASCE has recently updated its own history, *The American Civil Engineer, 1852-1974*, by William Wisely (1974). An encyclopedia edited by Antoine Picon, *L'art de l'ingénieur: constructeur, entrepreneur, inventeur* (1997), covers a broad range of structures, people, and ideas.

How do structures work? We recommended Mario Salvadori's books, including the classic *Structure in Architecture*, written with Robert Heller (3d ed., 1986). Rowland Mainstone in *Developments in Structural Form* (1975) explains structures through historical example; and Steven Vogel's *Cats' Paws and Catapults* (1998) is an intriguing exploration of technology, biology, and design. The illustrated books of David Macaulay will be familiar to many.

WHO'S RESPONSIBLE?

Abrams, Charles. *Forbidden Neighbors: A Study of Prejudice in Housing*. Port Washington, N.Y.: Kennikat Press, 1955, 1971.

American Society of Civil Engineers. *The Civil Engineer in Urban Planning and Development*. N.Y.: ASCE, 1966.

Asher, Robert L. "'Silicon Valley'—In Fairfax?," *Washington Post* (6 July 1985): A19.

Bowling, Kenneth R. *The Creation of Washington, D.C.: The Idea and Location of the American Capital*. Fairfax, Virginia: George Mason University Press, 1991.

Clines, Francis X. "Maryland Farmland: A Focus in Suburban Sprawl Battle." *New York Times* (25 June 2001): A10.

Fairfax County Economic Development Authority. *Fairfax County, Virginia, USA: Profile*. FCEDA, January 2001.

Garreau, Joel. *Edge City: Life on the New Frontier*. N.Y.: Doubleday, 1991.

Ginsberg, Steven. "Loudoun, Prince William Take a Unified Approach." *Washington Post* (23 December 2001): LZ01.

Jackson, Kenneth T. *Crabgrass Frontier: The Suburbanization of the United States*. N.Y: Oxford University Press, 1985.

Jackson, Richard J. and Chris Kochtitzky. *Creating a Healthy Environment: The Impact of the Built Environment on Public Health*. Washington, D.C.: Sprawl Watch Clearinghouse, 2001.

Jakle, John A. and Keith A. Sculle. *Fast Food: Roadside Restaurants in the Automobile Age*. Baltimore: Johns Hopkins University Press, 1999.

Johnson, George W. and John T. Hazel, "Virginia's Empty Coffers." *Washington Post* (5 November 2001): A23.

Kay, Jane Holtz. *Asphalt Nation: How the Automobile Took Over America, and How We Can Take It Back*. N.Y.: Crown, 1997.

Layton, Lyndsey. "Crowds Could Derail Decades of Progress." *Washington Post* (26 March 2001): A1.

Layton, Lyndsey. "Coming to a Curve; Regions' Subway System Begins to Show Its Age, Limits." *Washington Post* (25 March 2001): A1.

Lewis, David L. *Public Image of Henry Ford: An American Folk Hero and His Company*. Detroit: Wayne State University Press, 1976.

Lewis, Tom. *Divided Highways: Building the Interstate Highway, Transforming American Life*. N.Y.: Viking, 1997.

Marshall, Alex. *How Cities Work: Suburbs, Sprawl and The Roads Not Taken*. Austin: University of Texas Press, 2000.

Melton, R.H. "N. Va. Hopes to Share Wealth of Economy." *Washington Post* (21 December 1998): B1.

Metropolitan Washington Council of Governments. *Our Changing Region: Census 2000.* Publication 21810. Washington, DC: COG, 2001.

Monkkonen, Eric. *America Becomes Urban: The Development of U.S. Cities and Towns, 1780-1980.* Berkeley: University of California Press, 1988.

Ortega, Bob. *In Sam We Trust: The Untold Story of Sam Walton and How Wal-Mart is Destroying America.* N.Y.: Times Business, 1998.

"Prevost Hubbard." In U.S. Department of Transportation. *America's Highways, 1776-1976: A History of the Federal Aid Program* (Washington, D.C.: GPO, 1976): 322.

Schrag, Zachary M. "America's Subway: The Washington Metro as Vision and Vehicle, 1955-2001." Ph.D. diss., Columbia University, 2002.

Schrage, Michael. "Telemarketing Entrepreneur Keeps Trying; Latest Idea Fields Calls With Use of Computers." *Washington Post* (8 December 1986): F3.

Seely, Bruce E. *Building the American Highway System.* Philadephia: Temple University Press, 1987.

Seely, Bruce E. "Francis C. Turner: Father of the U.S. Interstate Highway System." *TR News* 213 (March-April 2001): 5-14.

Serwer, Andy and Julia Boorstin. "Ground Zero" [Northern Virginia], *Fortune* (9 October 2000).

Surface Transportation Policy Project. *Driven to Spend.* Washington, D.C.: STPP, 2000.

Swisher, Kara. *aol.com: How Steve Case Beat Bill Gates, Nailed the Netheads, and Made Millions in the War for the Web.* N.Y.: Three Rivers Press, 1998.

U.S. Department of Transportation, *America's Highways, 1776-1976: A History of the Federal Aid Program.* Washington, D.C: GPO, 1976.

Vance, Sandra S. and Roy V. Scott. *Wal-Mart: A History of Sam Walton's Retail Phenomenon.* N.Y.: Twayne Publishers, 1994.

Virginia Department of Transportation. *A History of Roads in Virginia.* Richmond: VDOT, 1989.

Wiseman, Lisa. "Easy Commuting, With a View" [Fairfax County]. *Washington Post* (22 December 2001): T5.

IS IT SAFE?

Appleyard, Donald. *Livable Streets.* Berkeley: University of California Press, 1981.

Bullard, Robert D., ed. *Sprawl City: Race, Politics, and Planning in Atlanta.* Washington, D.C.: Island Press, 2000.

Douglas, Mary and Aaron Wildavsky. *Risk and Culture: An Essay on the Selection of Technical and Environmental Dangers.* Berkeley: University of California Press, 1982.

Ewing, Reid. *Traffic Calming: State of the Practice.* Washington, D.C.: Institute of Transportation Engineers, 1999.

Fruin, John J. *Pedestrian Planning and Design.* Mobile: Elevator World, 1971, 1987.

Gladwell, Malcolm. "Wrong Turn," *The New Yorker* (11 June 2001).

Halsey, Maxwell. *Traffic Accidents and Congestion.* N.Y.: John Wiley, 1941.

Hass-Klau, Carmen. *The Pedestrian and City Traffic.* London: Belhaven Press, 1990.

Insurance Institute for Highway Safety. "Q & A: Pedestrians." www.iihs.org/safety_facts/qanda/peds.htm.

Institute for Transportation Engineers. *The Traffic Safety Toolbox: A Primer on Traffic Safety.* Washington, D.C.: Institute for Transportation Engineers, 1999.

Jacobs, Allan B. *The Boulevard Book: History, Evolution, Design of Multiway Boulevards.* Cambridge:, Mass. MIT Press, 2001.

Lalani, Nazir. *Alternative Treatments for At-Grade Pedestrian Crossings.* Washington, D.C.: Institute for Transportation Engineers, 2001.

McShane, Clay. *Down the Asphalt Path: The Automobile and the American City.* N.Y.: Columbia University Press, 1994.

Oglesby, Clarkson H. and Laurence I. Hewes, *Highway Engineering.* N.Y.: John Wiley, 1954, 1963.

"PEDNET, International Pedestrian Lexicon." user.itl.net/~wordcraf/lexicon.html.

Pline, James J., ed. *Traffic Engineering Handbook.* 4th ed. Englewood Cliffs, N.J.: Prentice Hall, 1992.

Portland Office of Transportation Engineering and Development. *Portland Pedestrian Master Plan.* June, 1998. *Portland Pedestrian Design Guide.* June 1998.

Pushkarev, Boris and Jeffrey M. Zupan. *Urban Space for Pedestrians: A Report of the Regional Plan Association.* Cambridge, Mass.: MIT Press, 1975.

Rudofsky, Bernard. *Streets for People: A Primer for Americans.* N.Y.: Doubleday, 1969.

Shinar, David. *Psychology on the Road: the Human Factor in Traffic Safety.* N.Y.: John Wiley, 1978.

Slovic, Paul, ed. *Perception of Risk.* London, UK: Earthscan Publications, 2000.

Southworth, Michael and Eran Ben-Joseph. *Streets and the Shaping of Towns and Cities.* N.Y.: McGraw-Hill, 1996.

Surface Transportation Policy Project. *Mean Streets 2000: Pedestrian Safety, Health, and Federal Transportation Spending.* www.transact.org.

U.S. Department of Transportation, Federal Highway Administration. *Speeding Counts On All Roads.* safety.fhwa.dot.gov/fourthlevel/pdf/Overview_Handout_1-42.pdf.

U.S. Department of Transportation, National Highway Traffic Safety Administration. *Traffic Safety Facts 1999: Pedestrians.* Washington, D.C.: National Center for Statistics and Analysis, 2000.

U.S. Department of Transportation, National Highway Traffic Safety Administration. *Traffic Safety Facts 2000.* www-nrd.nhtsa.dot.gov/pdf/nrd-30/NCSA/TSFAnn/TSF2000.pdf.

U.S. Department of Transportation, National Personal Transportation Survey. www.fhwa.dot.gov/ohim/nptspage.htm.

Zeeger, Charles, V., ed. *Design and Safety of Pedestrian Facilities.* Washington, D.C.: Institute of Transportation Engineers, 1998.

WHY SO BIG?

Campanelli, Ben. *Cellular Tower Guide.* Rochester, N.Y.: Benjamin F. Campanelli, 1997; rev. ed. 2000.

National Research Council. *In Our Own Backyard: Principles for Effective Improvement of the Nation's Infrastructure.* Washington, D.C.: National Academy Press, 1993.

National Association of Counties. *Building Together: Investing in Community Infrastructure.* National Association of Counties of the United States, 1990.

For public safety and risk, see Is It Safe? *above.*

IS IT AVAILABLE?

Galusha, Diane. *Liquid Assets: A History of New York City's Water System.* Fleischmanns, N.Y.: Purple Mountain Press, 1999.

Goldman, Joanne. *Building New York's Sewers: Developing Mechanisms of Urban Management.* West Lafayette, Indiana: Purdue University Press, 1977.

Granick, Harry. *Underneath New York.* N.Y.: Fordham University Press, 1991.

Griggs, Francis, ed. *A Biographical Dictionary of American Civil Engineers.* New York: American Society of Civil Engineers, 1991.

Hall, Edward. *Water for New York City.* Saugerties, N.Y.: Hope Farm Press, 1993.

Hardenbergh, W.A. *Water Supply and Purification.* 3d ed. Scranton: International Textbook Co, 1952.

Hopkins, Edward S. and Francis B. Elder. *The Practice of Sanitation.* Baltimore: Williams & Wilkins, 1951.

Jervis, John. *Description of the Croton Aqueduct.* N.Y.: Slamm and Guion, 1842.

Koeppel, Gerard T. *Water for Gotham: A History.* Princeton: Princeton University Press, 2000.

Melosi, Martin V. *The Sanitary City.* Baltimore: Johns Hopkins University Press, 2000.

Milwaukee Metropolitan Sewerage District. *Jones Island Wastewater Treatment Plant.* http://www.mmsd.com/about/page3.html.

Rogers, Peter. *America's Water.* Cambridge, Cambridge, Mass.: MIT Press, 1993.

Ryan, John C. and Alan Thein Durning. *Stuff: The Secret Lives of Everyday Things.* Seattle: Northwest Environment Watch, 1997.

Tarr, Joel A. *The Search for the Ultimate Sink.* Akron: University of Akron Press, 1996.

HOW MUCH DOES IT COST?

Barlow, John Perry. "A Declaration of the Independence of Cyberspace" (February, 1996). www.eff.org/~barlow/Declaration-Final.html.

Borsook, Paulina. "Real Time." *San Jose Mercury News* (27 February 2000).

Fisher, Jim. "Poison PCs." *Salon.com* (18 September 2000).

Goldberg, Carey. "Where Do Computers Go When They Die?" *New York Times*, (12 March 1998): G1/7.

Intel Corporation web site. www.intel.com.

Mann, Charles. "The End of Moore's Law?" *Technology Review* (May/June 2000).

Mieszkowski, Katherine. "Turn Off the Internet!" *Salon.com* (17 January 2001).

NSF/SRC Engineering Research Center for Environmentally Benign Semiconductor Manufacturing website. www.erc.arizona.edu.

"Putting It in Its Place," *The Economist* (11 August 2001): 18-20.

Reiterman, Tim. "San Franciscans Protest as Server Farms Sprout." *Los Angeles Times* (26 March 2001) A1/3.

Ryan, John C. and Alan Thein Durning. *Stuff: The Secret Lives of Everyday Things*. Seattle: Northwest Environment Watch, 1997.

Silicon Valley Toxics Coalition website. www.svtc.org.

Sweeney, Frank. "Projects Nurture Valley." *San Jose Mercury News* (21 December 1999).

Tajnai, Carolyn E. "Fred Terman, Father of Silicon Valley" (Stanford, May 1985). www.internetvalley.com/archives/mirrors/terman.html.

Zook, Matthew. "Connected as a Matter of Geography." *netWorker* 5/3 (2001): 13-17.

HOW LONG WILL IT LAST?

American Studies at University of Virginia 1930s Project. "Welcome to Tomorrow" [N.Y. World's Fair]. xroads.virginia.edu/g/1930s/DISPLAY/39wf/welcome.htm.

Augustyn, Robert T. and Paul E. Cohen. *Manhattan in Maps, 1527-1995*. N.Y.: Rizzoli, 1997.

Ayres, B. Drummond, Jr. "Sea Threatens Costly Building, Reviving a Debate." *New York Times*, (24 December 1997): A10.

"Backyard Burning" [PVC]. Maine Department of Environmental Protection. www.state.me.us/dep/air/backburn.htm.

Best, Don. "After the Storm: Hard-Won Lessons," *The Journal of Light Construction* (August 1993): 27-32.

Brand, Stewart. *How Buildings Learn: What Happens After They're Built*. N.Y.: Viking, 1994.

Buffalo Central Terminal from 1929 to the Present. BuffaloNet. terminal.BuffaloNet.org.

Buttelman, Michele E. "Historic Headlines: Remembering the St. Francis Dam Disaster." *The [Santa Clara] Signal* (11 March 2001).

Cruz, Laurence M. "Kingdome Imploded in Seattle." *Associated Press* (26 March 2000).

Duerksen, Chris and Robert Blanchard. *Belling the Box: Planning for Large-Scale Retail Stores*. Proceedings of the 1998 National Planning Conference.

Erlande-Brandenburg, Alain. *Notre-Dame de Paris*. Trans. John Goodman. N.Y.: Harry N. Abrams, 1998.

Feagans, Brian. "McIntyre Visits [Beach] Erosion Battle Zone," *Morning Star* (Wilmington, N.C.), April 12, 2001: 1A.

Fillion, Roger. "Railroaded" [fiber optic cable]. *The Industry Standard* (9 April 2001).

Findling, John E., ed. *Historical Dictionary of World's Fairs and Expositions, 1851-1988*. N.Y.: Greenwood Press, 1990.

Fortner, Brian. "High-Tech Highway," *Civil Engineering* (October 1999): 38-41.

Goldberg, Rob. "The Big Picture: Life Cycle Analysis." Philadelphia: Academy of Natural Sciences (May 1992). www.acnatsci.org/research/kye/big_picture.html.

Golden Gate Bridge Highway and Transportation District website. www.goldengatebridge.org.

Holding, Mark. *Staged Architecture: The Work of Mark Fisher*. London: Wiley-Academy, 2000.

Hodson, Jeff. "Built to Last: Next Week a Thing of the Past." *Seattle Times* (26 March 2000).

Kolle, Jefferson. "Weathering the Storm." *This Old House* (May 2001): 102-106.

Kronenburg, Robert. *Portable Architecture*. Boston: Architectural Press, 1996.

Leslie, Margaret. *Rivers in the Desert: William Mulholland and the Inventing of Los Angeles*. N.Y.: HarperCollins, 1993.

Levy, Matthys and Mario Salvadori. *Why Buildings Fall Down*. N.Y.: W. W. Norton, 1992.

Mair, George. *Bridge Down*. N.Y.: Stein and Day, 1982.

McGloin, John Bernard. *San Francisco: Story of a City*. San Rafael: Presidio Press, 1978.

McGrath, Gareth. "Pumping Sand Buys Time for Million-Dollar Beach Houses Facing an Inexorable Natural Cycle," [Wilmington, N.C.] *Sunday Star-News* (7 January 2001): 1A.

Monaghan, Frank, ed. *Official Guide Book, New York World's Fair, 1939*. N.Y.: Exposition Publications, 1939.

Mulholland, Catherine. *William Mulholland and the Rise of Los Angeles*. Berkeley, CA: University of California Press, 2000.

Smart Road web site. www.smartroad.org.

Steinberg, Ted. *Acts of God: The Unnatural History of Natural Disaster in America*. N.Y.: Oxford University Press, 2000.

Tanner, Jane. "New Life for Old Railroads" [fiber optic cable]. *New York Times* (6 May 2000): C1/4.

U.S. Department of Energy. "DOE's Yucca Mountain Studies," 1992.

Van Der Zee, John. *The Gate: The True Story of the Design and Construction of the Golden Gate Bridge*. N.Y.: Simon and Schuster, 1986.

Virginia Tech Transportation Institute website. www.vtti.vt.edu.

Wald, Matthew L. "Bury the Nation's Nuclear Waste in Nevada, Bush Says." *New York Times* (16 February 2002): A13.

Wald, Matthew L. "Energy Department Recommends Yucca Mountain for Nuclear Waste Burial." *New York Times* (15 February 2002): A19.

Walters, Jonathan. "Anti-Box Rebellion," *DBA Governing Magazine* (July 2000): 48.

White, John R. and Kevin D. Gray. *Shopping Centers and Other Retail Properties*. N.Y.: John Wiley, 1966.

Winston, Richard and Clara Winston. *Notre Dame de Paris*. N.Y.: Newsweek, 1971.

Wolf, Nancy and Ellen Feldman. *Plastics: America's Packaging Dilemma*. Washington, D.C.: Island Press, 1991.

Wurts, Richard and Stanley Appelbaum. *The New York World's Fair, 1939/1940 in 155 Photographs*. N.Y.: Dover Publications, 1977.

Zim, Larry et al. *The World of Tomorrow: the 1939 New York World's Fair*. N.Y.: Harper & Row, 1988.

Note: Articles available online often do not contain complete bibliographical information, thus we are unable to provide it.

Credits

Abbreviations: (t)-top; (b)-bottom; (r)-right; (l)-left; (c)-center; (g)- background; (i)-inset. Images not listed are courtesy of or copyright © 2002 Chicken&Egg Public Projects, Inc.

2, Mary Mahon. 6, (l) Collection of the New-York Historical Society; (lc) Ralph Bennitt, *Rolling Stones*, Evanston, IL: Row, Peterson, and Co., 1942; (c) Courtesy of Victor Ross and NYC DOT; (rc) Library of Congress. 7, (l) © Sears, Roebuck, and Co.; (c) Fairfax County Public Library; (rc) Library of Congress, Theodor Horydczak Collection, LC-H814-T-2921-x DLC; (r) Collection of the New-York Historical Society. 8, Library of Congress, Prints and Photographs Division, LC-USZ62-121881 DLC. 9, (bl) Jim Sichinolfi; (tr) Andy Craw/Washington County Communications (Oregon).

WHO'S RESPONSIBLE?

15, (l) Collection of the New-York Historical Society; (bc) and (br) Library of Congress, Prints and Photographs Division, Historic American Buildings Survey, HABS, VA, 30-2-7 and HABS, VA, 30-2-4. 17, thewheelmen.org and Wheelmen Richard Porath, Waltz, MI. 18, (l) and (tr) Library of Congress, Prints and Photographs Division, PAN US GEOG—Virginia, no. 20. 19, (l) Asphalt Institute Photo Archive; (b) Fairfax County Public Library. 20, (t) Collection of the New-York Historical Society. 22, (l) and (br) *TR News*. 24, (bl) Domenti-Foster Studios. 25, (tr) AP/Wide World Photos. 26, (l) and (bl) Wal-Mart Stores, Inc. 27, (l) Joe Alexander. 28, (l) Woodrow Wilson Center; (c) Dulles Corridor Rapid Transit Project; (r) Governor Glendening's Office. 29, (c) Trevor Wrayton/VDOT; (r) VDOT. 30, (t) Howard Chapin Ives, *Highway Curves*, John Wiley & Sons, Inc., 1956; (b) VDOT. 31, (l) *Washington Post*.

IS IT SAFE?

34, 36, 38, Jim Sichinolfi. 39, (l) TRAFFIC MANAGEMENT AND COLLISION by Clark. Reprinted by permission of Pearson Education, Inc., Upper Saddle River, NJ; (tr) and (br) Jim Sichinolfi. 40, (l) James Sichinolfi. (r) URS Corporation, Atlanta; (i) © ASTM. Reprinted with permission. 41, (l) Jim Sichinolfi; (tr) and (c) URS Corporation, Atlanta; (i) © ASTM. Reprinted with permission. 42, Jim Sichinolfi. 43, (l) ©1964 Institute of Transportation Engineers. Used by permission. 44, (l) American Automobile Association; (c) The Travelers Insurance Company, *Smash Hits of the Year*, 1940; (c) Courtesy of Victor Ross and NYC DOT. 45, The Travelers Insurance Company, *Smash Hits of the Year*, 1940. 46, Jim Sichinolfi; (i) *Life*, 24 September 1925. 47, (r) The Travelers Insurance Company, *Smash Hits of the Year*, 1940. 48, ©1956 SEPS: Licensed by Curtis Publishing Co., Indianapolis, IN, www.curtispublishing.com, artist: Steven Dohanos; (i) Department of Arkansas State Police, *Guide to Safe and Sane Driving*. 49, (l) The Travelers Insurance Company, *Smash Hits of the Year,* 1940; (r) TRAFFIC MANAGEMENT AND COLLISION by Clark. Reprinted by permission of Pearson Education, Inc., Upper Saddle River, NJ. 50, (l), 51, *Life*, 24 September 1925. 52, Collection of the New-York Historical Society. 53, (l) and (r) Courtesy of Victor Ross and NYC DOT; (c) Collection of the New-York Historical Society. 54, (tl) Museum of the City of New York; (bl) ©1998 Institute of Transportation Engineers. Used by permission. (c) Elevator World, Inc., 1987; (tr) Victor Ross/NYC DOT; (br) Bureau of Street Traffic Research. 55, (tl) Museum of the City of New York; all others courtesy of Victor Ross and NYC DOT. 56, (l) *Queens Courier*, (c) Courtesy of Victor Ross and NYC DOT. 57, (b) American Automobile Association. 59, (bl), 60, (l) and (r) City of Portland Office of Transportation, Pedestrian Transportation Program. 61, 62 (l), ©1999 Institute of Transportation Engineers. Used by permission.

WHY SO BIG?

67, (l), 68-73, David Arcos. 75, (tc), (tr), and second row (r) John Teague; second row (c) GATE/Global Academy of Tower Erectors; all others Anna Strout and Amy Wood. 80-81, Anna Strout and Amy Wood.

IS IT AVAILABLE?

89, (l) and (r) Collection of The New-York Historical Society. 90 (l) Jeffry Scott. 91, James Remick, Jr. Courtesy of the Jervis Library. 92, (all) Collection of The New-York Historical Society. 93, (t) and (bl) Collection of The New-York Historical Society; (br) New York Public Library. 94, (tr) and (br) Collection of The New-York Historical Society. 94-5, (g) New York Public Library. 95, (all) Collection of The New-York Historical Society. 96, Ted Dowey. 97, (l) *Transactions of the American Society of Civil Engineers*, 1913. 100, (t) and (b) © 2002 Public Works Journal. 102, Collection of The New-York Historical Society/Nicolino Calyo. 103, (l) Collection of The New-York Historical Society; (c) Collection of the NYC Fire Museum; (r) © 2002 Public Works Journal. 109, (cl) Downtown Collinsville, Inc., photo by Michael Gassmann; (cr) Board of Public Works, Gaffney, South Carolina. 111, Adam Hart-Davis. 112, (l) Collection of The New-York Historical Society; (tr) and (br) National Lime Association, www.lime.org.

HOW MUCH DOES IT COST?

116, Andy Craw/Washington County Communications (Oregon). 117, The historic collection of the City of Santa Clara, California. 118, (l) Joint Venture: Silicon Valley Network (www.jointventure.org); (t) and (b) The historic collection of the City of Santa Clara County, California; (c) Stanford University News Service. 120, (g) and (tr) The historic collection of the City of Santa Clara, California. 121, (t) University of Washington/Mary Levin; (c) Carnegie Mellon; (g) Varian Medical Systems. 122, Andy Craw/Washington County Communications (Oregon). 123, (r) World Gold Council. 124, (l) and (t) Intel Corporation; (br) Andy Craw/Washington County Communications (Oregon). 125, (l) and (r) California Department of Water Resources. 126, (all), 127, California Department of Water Resources. 128, (b) Intel Corporation; (c) NSF/SRC ERC for Environmentally Benign Semiconductor Manufacturing. 129, (l) MetroNexus; (r) Tonic 360. 130, (c) MetroNexus; (tr) and (br) Earthquake Protection Systems. 131, (tl) and (bl) MetroNexus; (r) Sacramento Bee/Chris Crewell. 133, (tl) MIT Museum. (bl) and (c) Parsons Brinckerhoff; (r) Wallace Engineering Structural Consultants. 135, (c) California Integrated Waste Management Board; (g) Silicon Valley Toxics Coalition. 136, Tom Carlson/Obsolete Computer Museum. 137, (t) and (c) HP Files/Hewlett-Packard Company; (b) California Integrated Waste Management Board.

HOW LONG WILL IT LAST?

142, (tr) Anna Strout; (br) Jim Sichinolfi. 144, Golden Gate Bridge, Highway and Transportation District. 145, (l) US Army Corps of Engineers; (t) and (c)Virginia Tech Transportation Institute; (b) Figg Engineering Group. 147, (r) Haskell reprint of the Castello Plan, 1660, Collection of The New-York Historical Society. 148, (c) Susan Dennis, (r) Edelman Worldwide. 149, (l) Ellerbe Becket. 150, (c) and (r) Collection of The New-York Historical Society. 151, (l) Collection of The New-York Historical Society; (r) Mark Fisher. 152, (bl) and (tr) NCDOT Photogrammetry Unit; (br) AP/Wide World Photos. 153, (bl) and (cr) AP/Wide World Photos; (tr) and (br) US Army Corps of Engineers. 154-5, (c) *Popular Science* magazine, ©1968 Times Mirror Magazines, Inc. 155, (bc) and (b) Los Angeles Public Library. 157, Library of Congress, Prints and Photographs Division, Historic American Buildings Survey, HABS, ALA, 63-TUSLO, 25-1. 158, (tl) and (bl) Steve Marcus/Las Vegas Sun. 158-9, Yucca Mountain Project. 160, (l), Library of Congress, Prints and Photographs Division, Historic American Engineering Record, HAER, NY, 30-HEMPN, 2-1; (c), Library of Congress, Prints and Photographs Division, Historic American Engineering Record, HAER, NY, 36-POJE, 2A-4; (r) Library of Congress, Prints and Photographs Division, Historic American Engineering Record, HAER, TEX, 155-WACO, 1-4.